LABORATORY MANUAL

MOE WASSERMAN

MICROELECTRONIC CIRCUITS AND DEVICES

SECOND EDITION

MARK N. HORENSTEIN

PRENTICE HALL, ENGLEWOOD CLIFFS, NJ 07632

Production Editor: *James Buckley*
Acquisitions Editor: *Alan Apt*
Supplement Acquisitions Editor: *Alice Dworkin*
Production Coordinator: *Julia Meehan*

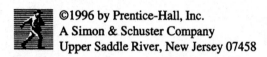
©1996 by Prentice-Hall, Inc.
A Simon & Schuster Company
Upper Saddle River, New Jersey 07458

All rights reserved. No part of this book may be
reproduced, in any form or by any means,
without permission in writing from the publisher.

Printed in the United States of America

10 9 8 7 6 5 4 3 2

ISBN 0-13-711182-9

Prentice-Hall International (UK) Limited, *London*
Prentice-Hall of Australia Pty. Limited, *Sydney*
Prentice-Hall Canada Inc., *Toronto*
Prentice-Hall Hispanoamericana, S.A., *Mexico*
Prentice-Hall of India Private Limited, *New Delhi*
Prentice-Hall of Japan, Inc., *Tokyo*
Simon & Schuster Asia Pte. Ltd., *Singapore*
Editora Prentice-Hall do Brasil, Ltda., *Rio de Janeiro*

TABLE OF CONTENTS

Experiment

	Guidelines for the Student	1
1	Linear Operational Amplifier Circuits	5
2	Nonlinear Operational Amplifier Circuits	9
3	Nonideal Operational Amplifiers	14
4	I-V Characteristics of the pn-Junction Diode	19
5	Diode Circuits	23
6	Small-Signal Characteristics of the Diode	26
7	Power-Supply Circuits	31
8	Zener-Diode Voltage Regulator	35
9	I-V Characteristics of the Metal-Oxide-Semiconductor Field-Effect Transistor (MOSFET)	38
10	I-V Characteristics of the Junction Field-Effect Transistor (JFET)	48
11	I-V Characteristics of the Bipolar Junction Transistor (BJT)	53
12	BJT and MOSFET Inverters	58
13	Transistor Biasing	63
14	BJT Common-Emitter Amplifier	69
15	MOSFET Common-Source Amplifier	75
16	BJT Emitter Follower	80
17	BJT Current Mirrors	84
18	BJT Differential Amplifier	88
19	Frequency Response of the Common-Emitter Amplifier	95
20	Transient Response of the pn-Junction Diode	99
21	Direct-Coupled Multistage Amplifier	104
22	Class AB Power-Amplifier Stage	108
23	Frequency Compensation of an Operational Amplifier	112
24	Comparison of Passive and Active Filters	119
25	Second-Order Active Filter	122
26	Oscillators	126
27	NMOS and CMOS Logic Gates	132
28	Propagation Delay of a CMOS Inverter	137
	Major Design Projects	142
	Appendix A: Use of the Breadboard	150
	Appendix B: Resistor Color Codes and Standard Values	151
	Appendix C: Use of the LM741 Operational Amplifier and Other Integrated Circuits	154

PREFACE

This Laboratory Manual is intended for use in elementary and intermediate courses in Electronics, Analog Circuits, and Digital Circuits. The sequence of topics follows closely the 2nd edition of *Microelectronic Circuits & Devices*, by Mark Horenstein. References to the applicable sections of that text are made at the beginning of each experiment, and the terminology and symbology follow that of the text. However, since material not covered in the Horenstein text is also included, and since the experiments are relevant to most courses in these subjects, it is hoped that the manual will be useful in conjunction with other texts as well.

Most laboratory manuals are designed to teach the student how to measure and analyze the properties of devices and circuits - certainly a necessary part of the educational process. However, since the ultimate task of many engineers is the design of circuits and systems, that aspect of the educational process is also addressed in this manual by the inclusion of design projects in many of the experiments. These may be used at the discretion of the instructor. An additional design component is included for those laboratory courses that may devote several weeks to a more comprehensive design project; several of the projects that have been used successfully at Boston University are provided in the final chapter. These require that the student do preliminary reading and planning before beginning the laboratory work. The longest of the projects will require four to five weeks to complete.

Each experiment begins with an <u>Introduction</u> that presents the concepts that underlie the experiment. The introductory material is intended as a supplement, rather than a substitute for the textbook material. The <u>Procedure</u> section often specifies work to be done by the student before coming to the laboratory. This not only reduces the time spent at the bench, but allows that time to be used more efficiently, since it encourages the student to plan the work beforehand. The student is not always given all the component values, voltages, and currents to be used in a circuit. This is done deliberately to encourage independent thinking. In most of the experiments there is an <u>Analysis of the Data</u> section that requires the student to perform calculations, to present the data in graphical form, and to evaluate the results critically. This section is omitted in experiments where it is more appropriate to ask for qualitative conclusions as part of the procedure. In these cases the questions are presented in *italic text*.

In some cases, related topics are included in one experiment. For example, Experiments 12 and 13 deal with both BJT and MOSFET inverters, and Experiment 25 includes both sinusoidal oscillators and the astable multivibrator. An attempt has been made to present the material in such a way that the instructor may select only those portions that are relevant to the course.

The first chapter, entitled <u>Guidelines for the Student,</u> contains information on general laboratory practice, circuit-board layout, troubleshooting technique, and current-measurement techniques. The student should be required to read this material carefully before beginning the first experiment. Appendixes A and B contain additional introductory material, which should also be read by the student before starting.

The student should have completed an introductory Linear Circuit Theory laboratory course as a prerequisite to the course that uses this manual. The new departure is, of course, the introduction of nonlinear circuit elements. Experience shows that two of the most difficult-to-grasp new ideas are the setting of an appropriate operating point for a nonlinear circuit and limiting the range of operation so that nonlinear elements can be represented by a linear small-signal model. Considerable introductory material has been devoted to these topics.

Many of the experiments in this manual have evolved from material used in courses at Boston University, to which many teachers have contributed. The contributions of Scott Dunham and Mark Horenstein have been especially valuable, and are gratefully acknowledged. My wife Josephine deserves special appreciation for her support and for proofreading the manuscript.

GUIDELINES FOR THE STUDENT

Use of This Manual

The introductiory material in each experiment is designed as a supplement, and not a replacement, for your reading in the course textbook. Both should be read before coming to the laboratory. In some experiments you will be asked to perform calculations or build a circuit ahead of time. This will allow you to use your laboratory time more efficiently.

This manual is not a "cookbook", with instructions to be followed uncritically. While performing an experiment, you may think of a "what if ...?" question that suggests another measurement, an extension of the range of measurement conditions, or even a circuit modification. You are encouraged to pose such questions, and to do whatever is necessary to answer them. There is no better way to learn the material.

You will often be asked to compare your experimental data with the results of calculations, and most of the time you will need resistance and/or capacitance values for these calculations. **Always use measured component values in your calculations. Do not depend on the nominal values, which may differ from the actual values by as much as 20 percent.** There are often legitimate reasons for discrepancies between your measured and calculated results, which you will be asked to discuss. To do this properly, you must first eliminate the spurious sources of discrepancies, such as incorrect component values.

General Laboratory Practice

You will most likely construct your circuits by plugging the components into a circuit board that requires no solder connections. The layout of one such board, the "Superstrip", is described in Appendix A to assist you in placing components. No matter how experienced one is in building circuits, the circuit will often not work the first time, because it was not connected properly, because one or more components are defective, because a wrong component value was chosen, or for some more obscure reason. Therefore there will always be the need to "troubleshoot" circuits by using meters or the oscilloscope to measure voltages and currents at various locations. For this reason, you should plan the layout of your circuit carefully so that troubleshooting can be done quickly and easily. The following guidelines are recommended:

1) always make a schematic diagram of the circuit before you begin to assemble components;

2) lay out the circuit on your board on a rectangular grid that resembles your schematic as closely as possible;

3) keep connecting wires as short as possible, so they will lie close to the board and not interfere with test probes;

4) do not connect the circuit to power supplies until it is completely assembled, and its layout has been checked with the schematic.

When you have to troubleshoot a circuit, the most efficient method is to separate the circuit into subsections whose behavior you can predict. First satisfy yourself that the input section performs properly. If component changes are needed, change only one component at at time, repeating your measurements until an incorrect or defective component, or an incorrect connection, is found. Once that section is operating properly, connect it to the next subsection and continue the procedure. Record your observations at each troubleshooting step; the record may be helpful if problems occur later.

Laboratory Safety

Although the low voltages and currents encountered in most of these experiments will present no hazard, you should still take reasonable precautions when handling electrical equipment. Special care is needed when you study power supplies, which are the only circuits that you will connect directly to the power line. Large-value electrolytic capacitors must be connected with the correct polarity shown on the case; if connected incorrectly they may explode and cause injury. Be sure to discharge capacitors before handling them.

Grounding Procedures

Strictly speaking, the only true ground point in the laboratory is a node connected to a copper rod driven into the earth, or a metallic cold-water pipe buried underground. The following points in the lab can be considered to be "more or less" at the same potential as this "system ground": the ground terminal on power supplies and signal generators; the frame of the oscilloscope; the third wire in the power cords. Note that the "low" terminal on a balanced-output signal generator is <u>not</u> at ground potential unless it is deliberately connected to one of the above points. It is good wiring practice to first connect together all "ground" terminals in a circuit, then to connect this common point to system ground through the oscilloscope frame, for example. All voltages in the circuit are then measured with respect to this common node.

When voltage and/or current is measured with two-terminal meters that have no ground terminal (usually true of battery-operated instruments), the voltmeter may be connected directly across the element being measured, and the ammeter may be placed anywhere in the loop whose current is being measured, without concern as to whether a node in the circuit is connected to the system ground through the 60-Hz power line. However, care must be taken when the oscilloscope or a three-terminal meter is used.

When an oscilloscope or a meter fitted with coaxial cable connectors is used to measure voltage, the outer conductor of the coaxial connector is permanently connected through the power line to system ground. Therefore one of the nodes of the circuit under test will be grounded directly. If another node is connected to system ground through some other path, at best the measurements will give incorrect results, and at worst there will be damage to circuit components.

An incorrect connection is shown in Fig. I-1, where the intention is to measure the voltage drop across R_1, given by $V_S\left(\dfrac{R_1}{R_1+R_2}\right)$.

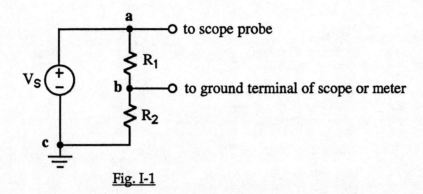

Fig. I-1

Node **c** is connected to system ground through the ground terminal of the voltage source V_S. Node **b** is connected to the same ground through the oscilloscope, with the result that resistor R_2 is short-circuited by a "ground loop". There is therefore no voltage drop across R_2. The oscilloscope measures the full voltage V_S across R_1 instead of the value expected from the voltage division expression. **The rule to remember is that no more than one node in the circuit may be connected to ground.** This circuit will give correct results if the voltage source is allowed to "float", with neither of its terminals connected to the ground terminal.

Current Measurement

In Experiment 1 you will measure voltage drops and currents in a circuit such as the one in Fig. I-2.

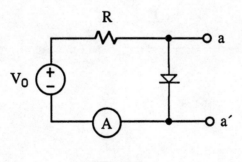

Fig. I-2

If you use a battery-operated milliammeter or microammeter, or one with both terminals ungrounded, you can inserted it as shown in the figure, whether or not there is a ground node in the circuit. However, current is often measured by using an oscilloscope to measure the voltage drop across a test resistor placed in the circuit branch of interest. Since most laboratory oscilloscopes have one input terminal permanently connected to the system ground, the arrangement of Fig. I-3 is recommended. Channel A of the oscilloscope measures the diode voltage, and channel B measures the voltage $-v_{test}$. The diode current i_D is v_{test}/R_{test} by Ohm's Law. If, for example, $R_{test} = 1\ k\Omega$, and the vertical sensitivity of the oscilloscope is

1 V/division, the oscilloscope has a current sensitivity of 1 mA/division. Note that the voltage drop across R_{test} is negative because the channel B input is negative relative to ground. Since the ground terminal in the test circuit is automatically determined by the connection to the oscilloscope ground, neither side of the voltage source may be grounded.

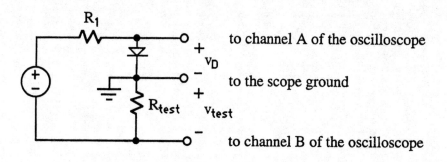

Fig. I-3

When the voltage source is a Tektronix FG503 function generator or similar instrument, which also has one of its terminals directly connected to system ground, the arrangement in Fig. I-4 may be used to avoid a ground loop. Since channel A now measures the diode voltage plus the drop across R_{test}, the resistor should be as small as possible, and the most sensitive range used for channel B, so the voltage drop across R_{test} can be neglected when compared with v_D. If that voltage drop cannot be made small enough, then the diode voltage must be obtained by subtracting the channel B reading from the channel A reading.

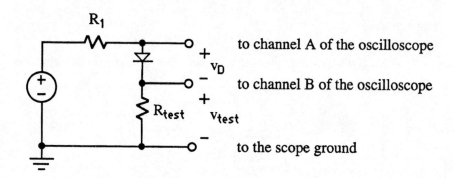

Fig. I-4

EXPERIMENT 1

LINEAR OPERATIONAL AMPLIFIER CIRCUITS

2.4 Linear Operational Amplifier Circuits
    2.4.1 Noninverting Amplifier
    2.4.3 Inverting Amplifier
    2.4.5 Op Amp Voltage Follower
    2.4.6 Difference Amplifier
    2.4.8 Summation Amplifier

Purpose

One of the most important uses of the operational amplifier (op amp) is in linear negative-feedback amplifiers with resistors in the feedback loop. In this experiment you will study the basic forms of this kind of amplifier.

Introduction

An operational amplifier, or op amp, shown symbolically in Fig. 1-1, provides an output voltage, referenced to ground, that is proportional to the difference between two input voltages, also referenced to ground.

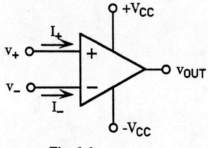

Fig. 1-1

Two of the most important characteristics of the op amp shown in Fig. 1-1 are

a) an extremely high <u>open-loop voltage gain</u> A_0, defined by

$$v_{OUT} = A_0(v_+ - v_-) \quad \text{for} \quad -V_{cc} < v_{OUT} < +V_{cc}, \tag{1-1}$$

and

b) $I_+ \approx I_- \approx 0.$ (1-2)

If the gain is sufficiently high, and the op amp operates in its linear region, then

$$v_+ - v_- = \frac{v_{OUT}}{A_0} \approx 0. \tag{1-3}$$

These features describe an <u>ideal op amp</u>, which we will use as our model in this experiment. Two additional properties of the ideal op amp are extremely high incremental input resistances, and essentially zero incremental output resistance.

The two basic feedback amplifier configurations are shown in Fig. 1-2.

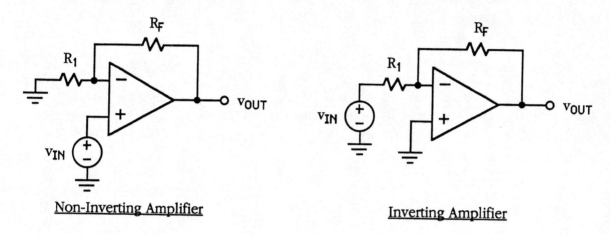

Non-Inverting Amplifier   Inverting Amplifier

Fig. 1-2

Note that the power supply terminals are not shown; this is usually done unless there is a special reason for showing them. Applying KCL at the inverting terminal node, and using (1-2) and (1-3) in both cases, gives

$$\frac{v_{OUT}}{v_{IN}} = 1 + \frac{R_F}{R_1} \quad \text{for the non-inverting amplifier} \tag{1-4}$$

and

$$\frac{v_{OUT}}{v_{IN}} = -\frac{R_F}{R_1} \quad \text{for the inverting amplifier.} \tag{1-5}$$

Many useful circuits are made by combining these two amplifier functions in various ways.

Procedure

I. Basic Inverting and Non-Inverting Amplifiers

1) Build the two amplifiers in Fig. 1-2. Use ±15-V power supplies for all the circuits in this experiment. If you have enough components, have both circuits available at the same time. Choose R_F and R_1 in the 1-kΩ to 50-kΩ range so that the ratio R_F/R_1 is the same for both circuits, and gives voltage gains around ±20. <u>Remember that the nominal resistor values shown on the color-coded bands may be significantly different from the actual values.</u> For v_{IN}, use a signal generator set to a 1-kHz sine wave, and adjust the amplitude to about 0.1 V p-p. Connect the signal generator to the input terminal of both amplifiers. Display v_{IN} and v_{OUT} for one of the amplifiers

on the oscilloscope, and measure the p-p amplitudes of both waveforms. Gradually increase the signal amplitude. *How do you explain what you see?* Repeat the procedure for the other amplifier. If you have difficulty obtaining such a small output from your signal generator, try using a larger output amplitude and a resistive voltage divider between the generator and the amplifier input to obtain the 0.1-V signal.

2) Now connect the signal generator to the x input of the scope and v_{OUT} of the non-inverting amplifier to the y input. Set the scope to x-y display mode, with an x sensitivity of about 0.5 V/div. You should now see a plot of the voltage transfer function v_{OUT}/v_{IN}. Vary the output voltage of the signal generator to see what happens. To get a better idea of how the plot is being made, set the generator frequency to about 0.1 Hz. Repeat this procedure for the inverting amplifier. *Describe and explain your observations in both cases.*

3) Connect an 0.1-V dc input to the inverting amplifier (you may need a voltage divider again), and measure the output voltages for a series of load resistors R_L connected between the output node and ground. Begin with $R_L = 100 \, \Omega$, and reduce the resistance until you reach the smallest value that is available to you.
Does anything unexpected happen to the output voltage as you reduce R_L? If so, can you explain it, using information on the manufacturer's data sheet for your op amp?

II. Voltage Follower

Build the circuit in Fig. 1-3, and perform the same operations that you did in Part I. *How does this circuit differ from the ones above? Of what practical use is it?* It is obviously related to the non-inverting amplifier. *Derive its voltage transfer function, starting with (1-4), and verify that your measurements are consistent with your calculated gain.*

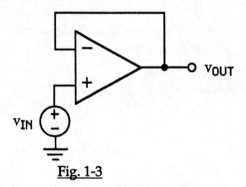

Fig. 1-3

III. Summing Amplifier

1) Build the circuit in Fig. 1-4, using 10-kΩ potentiometers. Pick $R_1 = R_2 = 10 \, k\Omega$, and $R_3 = 47 \, k\Omega$. With the power on, adjust the potentiometers until $v_1 = v_2 = 0$. Now experiment with several different combinations of potentiometer settings, measuring v_1, v_2, and v_{OUT} in each case. You should verify that the output voltage is the weighted sum of the two inputs, given by

$$v_{OUT} = -\left(\frac{R_3}{R_1}v_1 + \frac{R_3}{R_2}v_2\right). \tag{1-6}$$

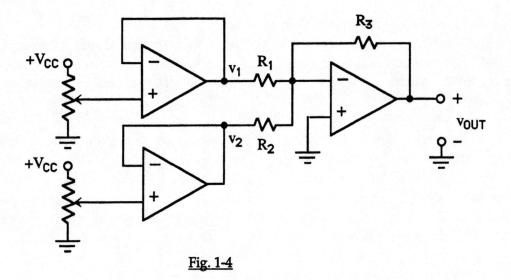

Fig. 1-4

How would the output voltage change if the unity-gain input buffers are removed without changing any resistor values or potentiometer settings?

2) Now remove the potentiometers and input buffers. Obtain two signal generators, connecting one to each of the input nodes v_1 and v_2. The two signals will be multiplied by their respective resistor ratios, then summed in accordance with (1-6). Display the output voltage on the oscilloscope. Experiment with signal generator outputs of different waveshape, amplitude, and frequency, and note the results. The more combinations you try, the more fun you will have. *Try to explain qualitatively as many of your observations as you can.*

IV. Difference Amplifier

Build the circuit if Fig. 1-5, using $R_1 = R_3 = 10$ kΩ, and $R_2 = R_4 = 47$ kΩ. <u>Measure all your resistor values with an ohmmeter; do not depend on the color codes</u>. Repeat the procedure of Part III-2 with this difference amplifier. *Derive the voltage transfer function (hint: use superposition of the two inputs), and compare your measured and predicted results.*

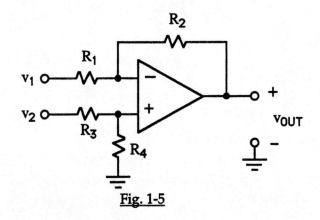

Fig. 1-5

EXPERIMENT 2

NONLINEAR OPERATIONAL AMPLIFIER CIRCUITS

2.5 *Nonlinear Operational Amplifier Circuits*
    2.5.1 *Open-Loop Comparator and Polarity Indicator*
    2.5.2 *Schmitt Trigger*

Purpose

Although op amps are most frequently used in linear applications that depend upon the presence of negative feedback, they have many nonlinear applications as well. In these cases they may be used without feedback (open-loop), or with various combinations of negative and positive feedback. A few of these nonlinear circuits will be studied in this experiment.

Introduction

Open-Loop Applications

Because the open-loop voltage gain of an op amp is so large, when there is no feedback, an input voltage difference of only a few microvolts is sufficient to drive the output voltage either to its maximum or to its minimum value, as determined by the supply voltages. This feature is used in comparator circuits, when one wishes to know whether a given input is larger or smaller than a reference value. Comparators are especially important in digital applications, such as in analog-to-digital converters.

Negative-Feedback Applications

Although feedback to the inverting input terminal allows the op amp to be used in many linear applications, it is possible to generate nonlinear input-output relationships by incorporating nonlinear elements in negative-feedback circuits. One of these circuits, the precision rectifier, or superdiode, will be studied here.

Positive-Feedback Applications

Feedback solely to the noninverting input terminal insures that the output voltage will be at one of its extreme values. A special class of comparators with memory can be made in this way. You will study the Schmitt Trigger in this experiment.

The Schmitt trigger (Fig. 2-1) is an extension of the comparator. The positive feedback and absence of negative feedback insure that the output will always be at either its maximum or its minimum possible value. The voltage divider R_1, R_2 sets v_+ at a fraction of the output. If $v_{IN} > v_+$, the output is negative, and if $v_{IN} < v_+$, the output is positive. Each time the difference $v_{IN} - v_+$

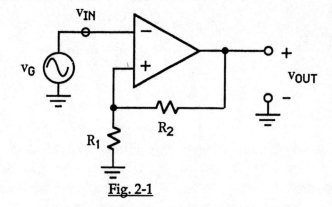

Fig. 2-1

changes sign, the polarity of the output, and consequently of v_+, changes. No further change is possible until v_{IN} reaches the new reference value v_+. The result is that the output may be at either extreme value for the same value of the input; whether the output is positive or negative is determined by its previous state. The circuit therefore possesses memory. The threshold for a change of output state is given by

$$v_{IN} > v_{OUT}\left(\frac{R_1}{R_1 + R_2}\right), \quad v_{OUT} < 0$$

$$v_{IN} < v_{OUT}\left(\frac{R_1}{R_1 + R_2}\right), \quad v_{OUT} > 0$$

(2-1)

Procedure

A. A Variable Open-Loop Comparator

1) Build the open-loop comparator circuit shown in Fig. 2-2, using ±15-V power supplies.

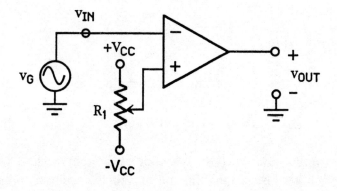

Fig. 2-2

R_1 should be a potentiometer in the 10-kΩ to 100-kΩ range. The purpose of R_1 is to set a variable reference voltage between +V_{CC} and -V_{CC} that can be applied as v_+ to the op amp. Use a digital voltmeter to measure v_{IN}, v_+, and v_{OUT}.

2) Set the signal generator to 1 kHz with an amplitude of the order of 1 volt. Record the value of v_+ for several positions of the potentiometer, and observe v_{OUT} on the oscilloscope in each case. Select both positive and negative values of v_+. Explain what is happening. Try increasing and decreasing the amplitude fo v_G, and *explain what you see*.

3) Display v_{OUT} vs v_{IN} using the x-y mode of the scope, The image will be the voltage transfer function of the comparator. Record its shape for several different settings of the potentiometer, and *explain what you see*.

B. The Precision Rectifier

You have learned that the simple diode rectifier with a resistive load gives an output voltage that increases approximately linearly with the input voltage, but only after the input exceeds a threshold of about 0.7 V, the "turn-on" voltage of the diode. When the input polarity is reversed, the output remains at zero because the diode is reverse-biased. The threshold can be reduced essentially to zero by adding an op amp to the circuit. This circuit is sometimes called a "superdiode".

1) Set up the two circuits of Fig. 2-3, and drive both simultaneously from one sinusoidal signal source that has about 3 V p-p amplitude. First, compare v_{OUT1} with v_G, using the dual-trace capability of the scope. Carefully reproduce the waveforms in your notebook. Then do the same with v_{OUT2}. *Explain the difference between the two outputs*. Remember that when D_2 is forward biased, the negative feedback loop is completed around the op amp, so that $v_+ \approx v_- = v_{OUT}$, and the circuit acts as a unity-gain amplifier. When D_2 is reverse-biased, the op amp functions in open-loop mode.

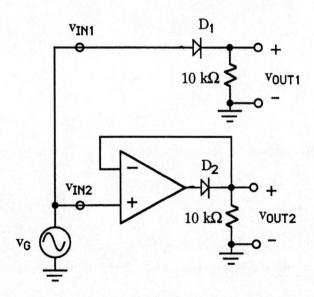

Fig. 2-3

2) Now reduce v_G to about 0.4 V p-p and compare the two output waveforms. *Does what you see make sense?*

C. <u>The Schmitt Trigger</u>

1) Build the circuit in Fig. 2-1, using ±15-V power supplies, and choosing R_1 and R_2 to be the same value in the range of 10 kΩ to 100 kΩ. Drive the circuit with a 1-kHz sine wave, beginning with an amplitude of a few tenths of a volt, and observe the output waveform. Try increasing the input amplitude, and record the observed output waveforms for different values of the input.

2) Connect the scope to display an x-y plot of v_{OUT} vs v_{IN}. Use an input frequency of about 0.1 Hz to allow you to follow the scope display. Vary the input amplitude and record your observations. *Explain what you see.*

Design Projects

A. Design, build, and test a Schmitt trigger that has the voltage transfer function shown in Fig. 2-4. Note that the hysteresis loop is not symmetrical about the vertical axis. This is accomplished by connecting feedback resistor R_1 not to ground, but to a dc reference voltage. Your first task is to find the design equations to give you values for the feedback resistors and the reference voltage.

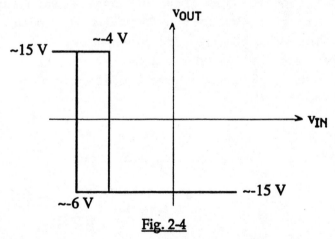

Fig. 2-4

B. Design, build, and test a negative-feedback circuit that generates an output voltage expressed by

$$v_{OUT} = V_1 \ln v_{IN} + V_2, \qquad (2-2)$$

where V_1 and V_2 are constants determined by the component values and characteristics. (Hint: whenever you see an exponential or logarithmic relationship, there is probably a diode in the circuit.)

C. Design, build and test a negative-feedback circuit that generates an output voltage expressed by

$$v_{OUT} = V_1 e^{\frac{v_{IN}}{V_2}}, \tag{2-3}$$

where V_1 and V_2 are constants determined by the component values and characteristics.

EXPERIMENT 3

NON-IDEAL OPERATIONAL AMPLIFIER CIRCUITS

2.6.2 Input and Output Offset Voltage
2.6.3 Input Bias and Input Offset Current
2.6.4 Slew Rate Limitation

Purpose

In all the experiments with op amps up to this point, it was assumed that no current enters the input terminals and that the output voltage is exactly zero when $v_+ = v_-$. This experiment deals with the real situation; you will measure how much error is incurred by making these approximations.

Introduction

Input Offset Voltage

When the input terminals of an op amp are connected together, the output voltage is generally not zero because of imperfect matching of components in the differential input stage of the amplifier. For example, if the two collector resistors in Fig. 18-3 (Experiment 18) are not identical, even if all other paired elements are exactly matched, there will be a voltage difference at the output terminals when the inputs are equal. The <u>input offset voltage</u> V_{IO} of the op amp is defined as the magnitude of the differential input voltage $v_+ - v_-$ that must be applied under open-loop conditions to set the output voltage to zero, or to overcome the internal offset. It is usually modeled as a positive voltage source connected to the non-inverting terminal of an otherwise ideal op amp. In the feedback amplifier of Fig. 3-1, the output voltage, in the absence of an external source, is

$$v_{OUT} = V_{IO}\left(1 + \frac{R_F}{R_1}\right). \tag{3-1}$$

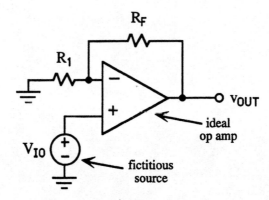

Fig. 3-1

Input Bias and Offset Current

In an op amp such as the LM741, which uses bipolar transistors as input elements, the base currents of the input transistors constitute the <u>input bias current</u>. The manufacturer's specification sheet cites a typical bias current of 200 nA. In an op amp such as the LF356, which uses JFET input elements, the input bias currents are far smaller because they are only the leakage currents of the reverse-biased gate junctions. With MOSFET input elements the bias currents are still smaller. If the input transistors were perfectly matched, the bias currents into each input terminal would be the same, and the <u>input offset current</u>, defined as the magnitude of the difference between the bias currents, would be zero. In practice, exact matching is impossible; therefore there is a non-zero input offset current.

The effect of input bias current can be found by analyzing Fig. 3-2. Since there is no external voltage source, there will be an output voltage IR_F caused by flow of the bias current through the feedback resistor.

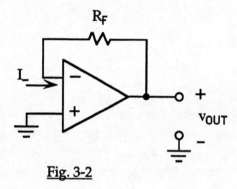

<u>Fig. 3-2</u>

Slew-Rate Limitation

The maximum rate, in volts/s or volts/µs, at which the output of an op amp can change is determined by the rate of charge or discharge of a capacitor inside the op amp. This quantity is called the <u>slew rate</u>. It is a large-signal quantity that is not directly related to the high-frequency rolloff point.. If, for example, the input signal to a linear feedback amplifier with voltage gain A_v is a sinusoid $V_A \sin\omega t$, then the slew rate limits the frequency and/or the amplitude of the output sinusoid $A_v V_A \sin\omega t$, above which there will be significant distortion of the waveform. The maximum rate of change is

$$\left.\frac{dv_{OUT}}{dt}\right|_{max} = \omega A_v V_A \cos\omega t \Big|_{max} = \omega A_v V_A < \text{slew rate} \qquad (3\text{-}2)$$

Any combination of the three factors that satisfies this condition will give an undistorted sinusoidal output. Fig. 3-3 shows what happens when the slew rate is exceeded. The dotted sinusoid is the expected output for a given amplifier and input signal 1. However, whenever the magnitude of the slope of the ideal output curve exceeds the slew rate, the output can change only at the slew rate. Therefore the ideal output curve (dotted sinusoid) is distorted into the slew-rate-limited output #1. The slew rate is the slope of the straight segments of this curve. Case #2 represents a smaller input signal or smaller gain. Since the slew rate is never exceeded, the ideal sinusoidal output (dashed sinusoid) is observed.

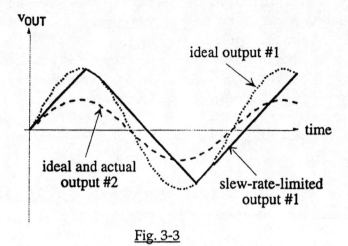

Fig. 3-3

Procedure

A. Input Offset Voltage

Build the circuit of Fig. 3-4. Choose the resistor ratio R_F/R_1 to give a measurable output voltage in response to the input offset voltage. You can estimate the value of V_{IO} from the manufacturer's data sheet for your op amp. Be sure that your closed-loop gain is much less than the open-loop gain of the op amp. If it is not, then the output will be saturated. Compare the data sheet value of V_{IO} with your value calculated from the output voltage.

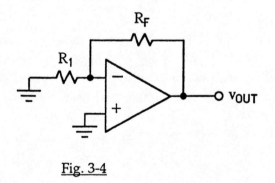

Fig. 3-4

B. Input Bias Current

Build the circuit of Fig. 3-2, using a value for R_F large enough to give a measurable output voltage due to the bias current. Remember that there will also be an output due to V_{IO}, but it will be small because this is a unity-gain amplifier with respect to V_{IO}. Calculate I_ from your measurement and compare it with the data-sheet value. Note that the data sheet gives a value for I_{BIAS}, which is the mean of I_+ and I_.

C. Offset Null

The output voltage due to the combined effects of offset voltage and bias current is a random quantity; it might even be zero in a particular circuit if the two effects happen to cancel each other. Many op amps provide an additional pair of <u>offset null</u> terminals that allow the designer to cancel out the unwanted output voltage by applying an external voltage via an external resistor. Although the nulling procedure is different for different types of op amp, many follow the procedure given below for the LM741. When in doubt, you must consult the manufacturer's data sheet for a given op amp type. The offset terminals for the LM741 are shown in Appendix C.

Build the non-inverting amplifier shown in Fig. 3-5, and set the input voltage to zero by grounding the input node.

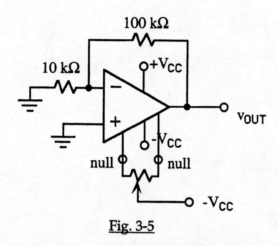

Fig. 3-5

Measure the output voltage with a digital voltmeter; the oscilloscope will not provide sufficiently accurate readings. Adjust the potentiometer setting and note its effect on v_{OUT}. *Can you find a setting that will give $v_{OUT} = 0$? What is the full range over which you can change v_{OUT} by adjusting the null control?*

D. Slew Rate

Build the unity-gain follower of Fig. 3-6, using ±15-V power supplies, and set the input source to a 100-Hz square wave with 10-V p-p amplitude. At this amplitude the output will not saturate. Gradually increase the frequency of the input square wave until the output ceases to look like a square wave on the scope trace; at this point it is slew-rate limited. If a Polaroid camera is available, position the trace in the upper half of the scope screen and photograph it. Do not develop the photograph yet. Your goal is to obtain two scope traces on the same print.

Now, without changing the input signal frequency, switch to a 10-V p-p sine wave Position the trace of v_{OUT} in the lower half of the screen and take a second exposure. Paste the photograph in your notebook and label the two traces. If a camera is not available, then sketch the two waveforms in your notebook as carefully as you can.

Finally, switch the generator back to a square wave, and increase the frequency until the output looks like a triangular wave. The op amp is now strongly slew-rate limited. Measure the slew rate directly from the slope of the waveform, and compare it with the value given in the data sheet for the op amp.

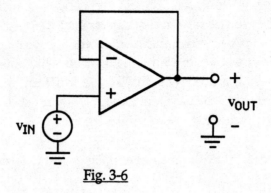

Fig. 3-6

EXPERIMENT NO. 4

I-V CHARACTERISTICS OF THE SILICON PN-JUNCTION DIODE

3.3 Examples of Two-Terminal Nonlinear Devices
  3.3.1 Semiconductor Materials
  3.3.2 Physical Characteristics of the PN Junction Diode
  3.3.3 Voltage-Current Characteristic of the PN Junction Diode

Purpose

To study the current-voltage relationship of a <u>nonlinear circuit element</u>, the silicon pn-junction diode. The diode is an important circuit element in its own right, and it is also an essential part of the bipolar junction transistor.

Introduction

A silicon pn-junction diode is made by introducing acceptors, or p-type dopants into one region of a silicon crystal, and donors, or n-type dopants into an adjacent region. The p-type side contains many more free positive holes than free negative electrons, and the n-type side has the opposite distribution. Whenever there is a nonuniform distribution of particles of any kind (charged or uncharged) in any system, there will be a net motion of the particles by <u>diffusion</u> from the high-concentration region to the low-concentration region. In the doped silicon the motion of these <u>carriers</u> will constitute an electric current because the particles are charged. The hole and electron currents will add because of the opposite charges and opposite directions of motion.

If the silicon crystal is not connected to an external circuit, there can be no net current flow in spite of the diffusion process just described. Therefore there must be an opposing mechanism that exactly cancels the <u>diffusion current</u>. This <u>drift current</u> arises from the electric field that forms at the pn junction when the carriers move into the opposite region, leaving behind immobile acceptor and donor atoms, that were originally electrically neutral, with a net charge opposite to that of the carrier that moved away. The junction diode is shown in Fig. 4-1, with the four current components depicted by arrows. The width of the <u>depletion layer</u>, which is devoid of free carriers, and contains only ionized donor and acceptor atoms (shown by the positive and negative sign symbols), is not drawn to scale; it is actually very narrow compared with the overall dimensions of the diode.

When the diode is connected to a voltage source with the p-type side positive, the electric field inside the silicon is reduced because the direction of the field component introduced by the external source is opposite to that of the internal field. The drift component of the current is thus reduced, and it is no longer large enough to cancel the diffusion component exactly. This imbalance results in a net flow of diffusion current in the direction determined by the external voltage source. When the polarity of the external source is reversed, the drift component dominates. However, this component is negligible because the concentrations of minority-carrier free holes on the n side and minority-carrier free electrons on the p side are extremely low. The diode current i_D is thus an asymmetric function of the external voltage v_D; it is given by the exponential function in equation (4-1):

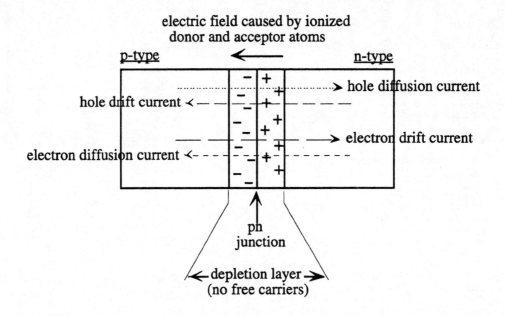

Fig. 4-1

$$i_D = I_s \left(e^{\frac{v_D}{\eta V_T}} - 1 \right) \tag{4-1}$$

where I_s and η are constants for a specific diode, and V_T is the thermal voltage kT/q, which equals 0.026 V at room temperature. In this experiment the i-v characteristic of the 1N4148 semiconductor diode, or its equivalent, will be examined in detail and compared with the above equation. The values of the two diode constants will be determined from the data.

Procedure

The circuit of Fig. 4-2 is designed to analyze the forward-biased portion of the diode i-v characteristic by applying a load line of constant slope but varying open-circuit voltage to the diode. Note that R_1, which is equal to 1 kΩ, safely limits the current through the diode so that it won't burn out when V_0 becomes large. It is important that V_0 not be too high, to insure that both the diode and R_1 operate within their power limitations.

1) Construct the circuit of Fig. 4-2, and measure the diode voltage v_D between the nodes **a** and **a'** with the oscilloscope set to the 0.1 V/div scale, and the diode current i_D with a milliammeter

or microammeter in series (represented by the encircled **A**). See the first chapter, Guidelines for the Student, for a discussion of current measurement techniques.

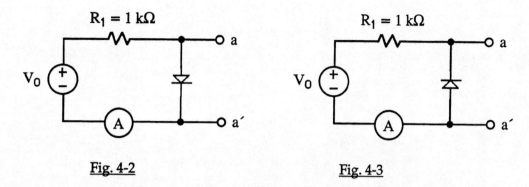

Fig. 4-2   Fig. 4-3

Begin with V_0 set to zero (open-circuit Thevenin voltage at the a-a' terminals equal to zero), and gradually increase it, so that different diode operating points are established. Record v_D for several values of i_D up to at least 10 mA. Be sure to record at least ten data points in the range below about 200 µA. Note that you are only interested in the diode voltage v_D, and not the voltage V_0 applied to the entire circuit.

2) Return V_0 to zero, and rearrange the circuit as shown in Fig. 4-3, which is designed to measure the reverse-bias diode characteristic. As indicated by equation (4-1), minimal current flow can be expected. Set the milliammeter or microammeter to its lowest scale to measure the current.

Begin with V_0 at zero, and gradually increase the reverse voltage applied to the diode, recording the current and voltage at several points. *Does your i-v characteristic seem to have the general form expected? If not, how might you modify the circuit of Fig. 4-3 to make the measurement more accurate?*

Analysis of the Data

A) Plot i_D vs v_D from parts 1 and 2 on the same piece of engineering graph paper. Estimate from the graph the threshold, or turn-on voltage V_f for currents approximately in the 1-10 mA range, and also estimate the reverse saturation current I_s.

B) On another piece of graph paper, plot i_D vs v_D for the current range from zero up to about 200 µA. *What do you conclude about the significance of the turn-on voltage V_f? Is it a constant for a given diode?*

When the diode is forward biased, the exponential term in (4-1) is so large that the -1 term can be neglected. Then, by taking logarithms of both sides, we get

$$\ln i_D \approx \frac{v_D}{\eta V_T} - \ln I_s \quad \text{or} \quad \log_{10} i_D \approx \frac{v_D}{2.3\eta V_T} - \log I_s \qquad (4\text{-}2)$$

C) Make a plot of log i_D vs v_D in the forward-bias region on semilog graph paper. Given the thermal voltage $V_T = 26$ mV at room temperature, determine the value of η and a more precise value for I_s.

D) Label the diode whose parameters you have just measured so you can use it in later experiments.

EXPERIMENT NO. 5

DIODE CLIPPING AND LIMITING CIRCUITS

4. *Signal Processing and Conditioning with Two-Terminal Nonlinear Devices*
    4.2 Clipping and Limiting Circuits
    4.3 Rectifier Circuits

Purpose

To use the pn-junction diode in circuits in which the output signal is not a linear function of the input signal. In particular, you will study <u>limiter</u>, or <u>clipper</u> circuits, in which the output is a function of the input in certain ranges, and essentially independent of the input in other ranges of operation.

Introduction

In many circuits where diodes are employed, the constant-voltage-drop approximation, or battery-ideal diode model (Fig. 5-1) is sufficient for describing the diode behavior. On the basis of the results of Experiment 4, you should formulate such a model for your 1N4148 diode that applies in the 1-10 mA range, and use it to analyze the diode circuits in this experiment.

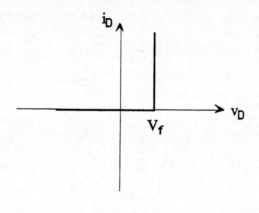

Fig. 5-1

In a clipping circuit, the output voltage will be proportional to the input voltage as long as the input lies between specified reference levels. Outside this range the output is "clipped"; it remains essentially constant, no longer dependent on the input. Clipping circuits find important uses in wave-shaping and signal-processing applications. Several circuits of this and related types will be studied in this experiment.

Procedure

1) Construct the circuit of Fig. 5-2 and set the input voltage v_s to a sine wave of at least 1 volt peak amplitude. Try changing the amplitude of the input signal. *Based on the your diode model, how do you explain the output waveform? Why is the sine wave not clipped exactly flat?* Make measurements over a sufficiently large range of inputs so that you can plot the voltage transfer characteristic of this circuit.

2) Add the dc voltage source V_1 shown in Fig. 5-3 to the circuit, and set the input signal to a triangular wave of 20 volts peak-to-peak. Slowly vary V_1 over the range 0 to 15 volts. Observe the output waveform. *Does it agree with what you predict from your model?*

3) Add a second diode and voltage source V_2 (Fig. 5-4). <u>Connect V_2 with the polarity shown, or you may burn out both diodes.</u> (*Why?*) Experiment with different levels of both voltage sources, make sketches, and comment. This circuit is similar to the limiting circuit used in FM radio receivers.

4) Set V_1 to about 8 volts and reverse the polarity of V_2 to a positive voltage of +5 volts. V_1 and V_2 should now have the same polarity. (Note: <u>V_2 must be less than V_1, or you will burn out the diodes. *Why?*</u>) Comment on the observed waveform. Draw the voltage transfer characteristic of this circuit.

5) Construct the half-wave rectifier clipper circuit of Fig. 5-5, and try a few values of V_1. Comment on the observed output waveform. Draw the voltage transfer characteristic.

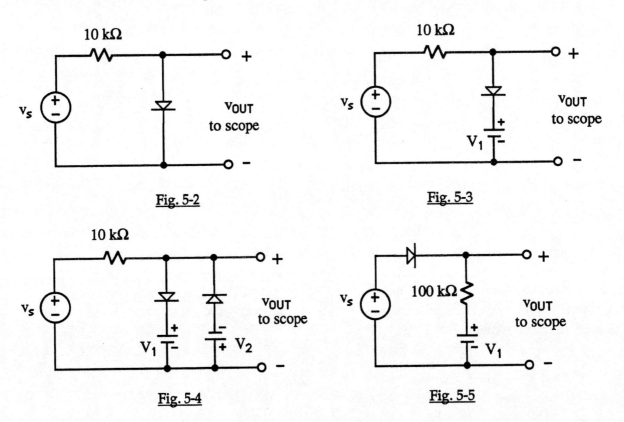

Fig. 5-2   Fig. 5-3

Fig. 5-4   Fig. 5-5

Design Projects

The networks described below may contain no components other than voltage sources, diodes, and resistors.

A. Design, build, and test a network that has an i-v characteristic approximately as shown below in both graphic and tabular form:

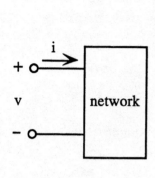
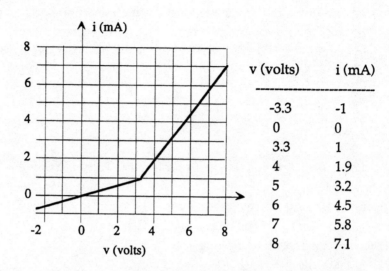

v (volts)	i (mA)
-3.3	-1
0	0
3.3	1
4	1.9
5	3.2
6	4.5
7	5.8
8	7.1

B. Design, build, and test a network that has a voltage transfer characteristic approximately as shown below. Note that this is a v_{OUT} - v_{IN} characteristic, not an i-v characteristic.

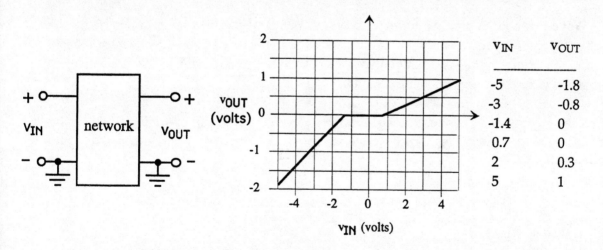

v_{IN}	v_{OUT}
-5	-1.8
-3	-0.8
-1.4	0
0.7	0
2	0.3
5	1

EXPERIMENT 6

THE SMALL-SIGNAL MODEL OF THE DIODE

Purpose

To determine the conditions under which the diode acts as a linear element, in which a <u>change</u> in voltage causes a proportionate <u>change</u> in current. This is called <u>small-signal</u>, or <u>incremental</u> operation. Small-signal analysis will also be used later when you study the voltage gain of a transistor amplifier.

Introduction

The constant-voltage-drop, or battery model of the diode is obviously an approximation. Deviations from it became apparent in Experiment 5, when you observed that the output voltage of a clipping circuit is not truly independent of the input voltage. The purpose of this experiment is to look more closely at the diode circuits, and to interpret their behavior with a more precise model. We return to the exponential diode equation:

$$i_D = I_s \left(e^{\frac{v_D}{\eta V_T}} - 1 \right) \tag{6-1}$$

When the diode is placed into forward bias at an <u>operating point</u> determined by the dc quantities V_D and I_D, a small variation around this point in the diode current will cause a small variation in the diode voltage. Although this voltage variation is not predicted by the battery model, it is explained by the exponential model.

The response of the diode to small variations in voltage about an operating point can be described by the <u>incremental resistance</u> r_d, defined by the derivative dv_D/di_D, evaluated at the operating point:

$$\frac{1}{r_d} = \frac{di_D}{dv_D}\bigg|_{op.pt.} = \frac{I_s}{\eta V_T} e^{\frac{v_D}{\eta V_T}}\bigg|_{op.pt.} = \frac{I_s}{\eta V_T} e^{\frac{V_D}{\eta V_T}} \approx \frac{I_D}{\eta V_T} \tag{6-2}$$

The approximation is valid in forward bias, where the -1 term in the current equation is negligible compared with the huge exponential term. The smaller the value of r_d, the more steeply rising the i-v curve. The battery model implies $r_d = 0$. For a given operating-point current I_D, the incremental resistance is constant. Therefore the change in current about the operating point is proportional to the change in voltage. This can be expressed as a form of Ohm's Law:

$$\Delta v_D = r_d \Delta i_D \quad \text{or} \quad v_d = r_d i_d \tag{6-3}$$

where the value r_d depends upon the operating point. Stated another way, the exponential i-v curve of the diode can be approximated by a straight line over a small portion of the curve. This approximation is shown graphically in Fig. 6-1, where a small part of the exponential i_D vs v_D curve is plotted for a typical diode.

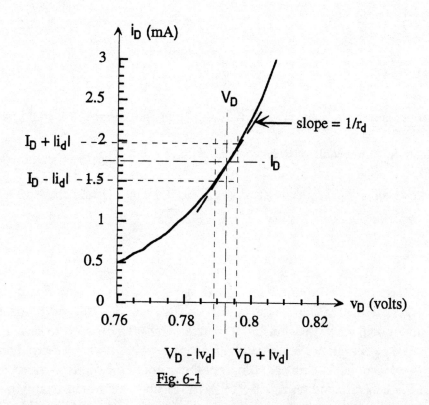

Fig. 6-1

Note the terminology used here. A lowercase symbol with uppercase subscript (v_D) represents the total value of a quantity. An uppercase symbol with uppercase subscript (V_D) represents the dc, or average, or operating-point value. A lowercase symbol with lowercase subscript (v_d) represents the incremental value, or deviation from the dc value. The deviation may be positive or negative, even if the total value always has the same sign. For any quantity x, which may be time-dependent, we can always write

$$x_A(t) = X_A + x_a(t) \tag{6-4}$$

or the total value is the sum of the dc component and the incremental component.

The answer to the question of how small v_d must be for the small-signal approximation to hold can be answered by deriving (6-2) in a different way. Starting with (6-1) and expressing the total diode current and voltage as in (6-4), we get, for forward bias,

$$I_D + i_d \approx I_s e^{\frac{(V_D + v_d)}{\eta V_T}} = I_s e^{\frac{V_D}{\eta V_T}} e^{\frac{v_d}{\eta V_T}} = I_D e^{\frac{v_d}{\eta V_T}} \tag{6-5}$$

Now the exponential term is replaced by the series expansion

$$e^x = 1 + x + \frac{x^2}{2!} + \frac{x^3}{3!} + \dots \quad \text{where} \quad x = \frac{v_d}{\eta V_T} \qquad (6\text{-}6)$$

and the higher power terms are neglected by making $x \ll 1$, or, in our case, $v_d \ll \eta V_T$. Equation (6-5) then becomes

$$I_D + i_d \approx I_D\left(1 + \frac{v_d}{\eta V_T}\right) \quad \text{or} \quad i_d = \frac{I_D v_d}{\eta V_T} \qquad (6\text{-}7)$$

This is the same result as (6-2), but the derivation shows that the result is valid only if

$$v_d \ll \eta V_T \approx 26 \text{ mV at room temperature, since } \eta \approx 1. \qquad (6\text{-}8)$$

In practice, the small-signal approximation begins to fail if v_d exceeds about 10 mV.

Procedure

1) Construct the circuit of Fig. 6-2, using a dc source with floating output terminals for V_S and a 1-kHz sinusoidal source for v_g. Choose R_y around 100 Ω and R_x around 1 kΩ. This voltage attenuator will allow the sinusoidal voltage across the diode v_s to be as little as a few millivolts when the generator v_g is operated on a higher voltage scale, where more accurate control of its output is achievable. The dc operating-point current of the diode is measured with an ammeter that has no grounded terminals. The oscilloscope will measure only the sinusoidal components v_d of the diode voltage and v_s of the input voltage because the coupling capacitors block the dc components. Do not use the 10x oscilloscope probes for these measurements, so that the scope will operate at maximum vertical sensitivity.

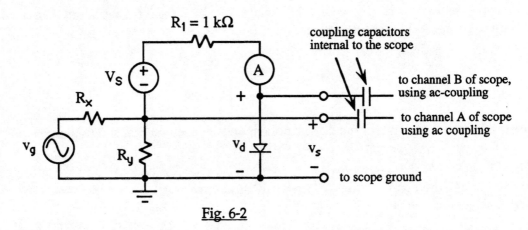

Fig. 6-2

2) With v_g set to zero, choose a value of V_S to place the diode in forward bias with a current of about 1 mA. Record the current. After adjusting the baselines of the two oscilloscope traces so that they coincide, increase v_g until v_s is about 5 mV. Then, with the gain control of channel B at or near its most sensitive <u>calibrated</u> position (probably 2 mV/division or 5 mV/division), adjust the gain control of channel A so that the sinusoidal traces of v_s and v_d are superimposed as closely as you can. You will have to use an uncalibrated scale factor for channel A to do this. Now increase

v_s slowly, observing the oscilloscope traces. Record as precisely as you can the value of v_d if and when the two traces begin to diverge. It should not be necessary for v_s to exceed a few hundred millivolts to reach this point. The value of v_s is not important.

3) Return v_s to zero, and return the gain control of channel A to a calibrated position. Set the oscilloscope to the x-y display mode, with v_s on the x axis and v_d on the y axis. It will be convenient to set the origin at the center of the oscilloscope display. Be sure that the inputs are still ac coupled. Increase v_s again, and reproduce the observed x-y plot in your notebook as accurately as you can. A Polaroid photograph of the display is preferable if a camera is available.

4) Repeat steps 2 and 3 after setting a new value of V_S that gives a dc diode current of about 10 mA.

Analysis of the Data

If the incremental resistance r_d of the diode is constant, the v_d vs v_s plots should be straight lines. The values of r_d can be obtained from the slopes of the lines by using the voltage divider expression

$$v_d = v_s \left(\frac{r_d}{r_d + R_1} \right). \tag{6-9}$$

This expression follows from the incremental circuit in Fig. 6-3. If you find that the uncertainty in the calculation is too great, try repeating the measurements with a smaller value of R_1. It will be necessary to use correspondingly smaller values of V_S to keep the total current in a safe range.

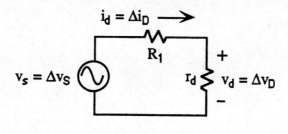

Fig. 6-3

Note that the dc quantities V_S, I_D, and V_D do not appear in this circuit - only the changes, or increments, in the total quantities. Compare the values of r_d obtained this way with the values you calculate from equation (6-2) for each value of I_D.

Are your voltage-transfer plots straight lines over their entire range? Can you explain any deviations from linearity that you might observe? Compare your observations with the result expected from equation (6-7).

Design Project

Design, build, and evaluate a <u>voltage-controlled attenuator</u> for sinusoidal signals. The external configuration and the approximate relationship between the attenuation factor and the control voltage are shown in Fig. 6-4. The circuit inside the shaded rectangle may contain any combination of sources, diodes, resistors, and capacitors.

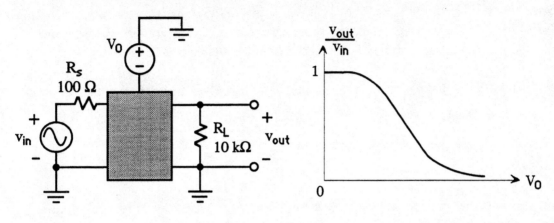

Fig. 6-4

For $v_{in} = V_A \sin\omega_1 t$ (dc component = 0), you should measure $v_{out} = \alpha V_A \sin\omega_1 t$ (dc component = 0). The attenuation factor α, or ratio of output amplitude to input amplitude, should be less than unity, but should be close to unity when the control voltage $V_0 = 0$. No scales are shown on the axes because you may choose them yourself. Explain the operation of your circuit.

EXPERIMENT 7

DC POWER SUPPLY CIRCUITS

4.4 Power Supply Circuits
   4.4.1 Half-Wave Rectifier Power Supply
   4.4.2 Full-Wave Rectifier Power Supply

Purpose

To study elementary <u>dc power supply circuits</u>, which transform the energy from the 60-Hz, 120-V power line to a steady dc that can be used by electronic circuits and equipment.

Introduction

In almost every dc power supply, a rectifier circuit and capacitor filter are used to convert an ac supply voltage to a dc voltage and to smooth out the dc level over time. Because most electronic circuits require dc power, rectifier circuits are almost always found in electronic equipment that operates from the 60-Hz ac power lines. Although sophisticated voltage-regulating circuits are usually employed, the rectifier circuits described in this experiment form the basis from which commercial dc power supplies are made.

You have already studied the simple diode rectifier in experiment 5 (Fig. 5-5, with V_1 set to zero). Although that circuit converted an ac input to an output that is never negative, the output amplitude varies from zero almost to the peak value of the input. To be useful, a power supply must maintain an output voltage that is constant over time, and that does not change when the input voltage or the load resistance changes. This experiment and the one that follows are devoted to the accomplishment of these goals.

The essential part of a dc power supply is the series diode-capacitor structure in Fig. 7-1.

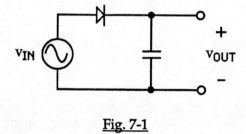

Fig. 7-1

The capacitor charges during the half cycle when the diode is forward biased, and retains its charge during the other half cycle. The output voltage is thus constant, with peak value approximately 0.7 V below the peak of v_{IN}. Of course, a load must be connected if this circuit is to do any useful work. Unfortunately, as soon as the load is connected, v_{OUT} will no longer be constant, because the capacitor begins to discharge through the resistor.

Procedure

Half-Wave-Rectifier Power Supply:

1) The circuit in Fig. 7-2 is the simplest of all power supply rectifier circuits. In this case a 60-Hz ac transformer provides the input voltage to the rectifier circuit. Construct the circuit of Fig. 7-2, at first omitting the filter capacitor. (Note that until the capacitor is added, the half-wave rectifier falls into the clipper circuit family.) Also omit at first the 100-ohm load resistor. But include the 100 kΩ resistor so that the circuit has some minimal load.

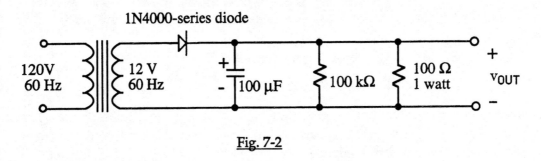

Fig. 7-2

Observe the output waveform on the scope. What you see is often referred to as "pulsating dc" in power-supply language. What is the zero-to peak magnitude of this waveform? *Does it agree with what you would expect, given the voltage specifications of your transformer? Explain any seeming discrepancies; everything should make sense before you proceed.*

2) Now add the filter capacitor to the circuit, as shown in Fig. 7-2. Be sure to connect the electrolytic capacitor with proper polarity. It may explode if you don't!! Observe the output waveform on the scope, and also measure it with a voltmeter. Now add the 100-Ω, 1-watt resistor and observe and measure again. Be sure that you can explain what you see, both with and without the load resistor. *Can you explain the ripple voltage that you see?*

Measure the percent ripple with a 100-Ω load, then estimate the value of capacitance required to achieve less than 5% ripple with the 100-Ω load. Are you able to find capacitors large enough to achieve this level of ripple? If not, assemble as much capacitance as you can by combining units in parallel. Calculate the amount of ripple they will produce, install them, and compare your observations with your prediction. Is this supply a very well filtered one? Use this equation to calculate the ripple:

$$\% \text{ ripple} = \frac{\text{p-p ripple} \times 100}{\text{average dc}} \qquad (7\text{-}1)$$

These quantities are defined in Fig. 7-3.

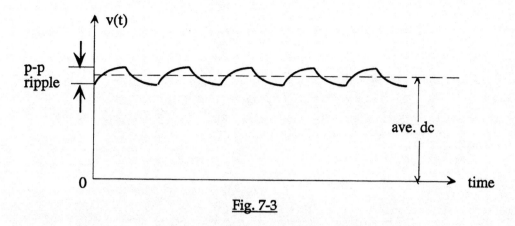

Fig. 7-3

The regulation of a power supply is a measure of how much the output voltage changes as more and more current is drawn from the output terminals. Percent regulation is defined by:

$$\% \text{ regulation} = \frac{\langle \text{no-load voltage} \rangle - \langle \text{full-load voltage} \rangle}{\langle \text{full-load voltage} \rangle} \qquad (7\text{-}2)$$

where $\langle .. \rangle$ denotes the time-average dc quantity. Note that the ideal zero-percent ripple is achieved when the supply is perfectly regulated. What is the percent regulation of the half-wave-rectifier power supply of Fig. 7-2, assuming that the 100-Ω resistor constitutes a full load?

Full-Wave Bridge Rectifier

3) Construct the bridge rectifier of Fig. 7-4, and observe the output waveform without the filter capacitor and load resistor. *How does this pulsating dc waveform compare with that of the half-wave supply?* Now add the filter capacitor and the 100-Ω load resistor. *Is the ripple improved over that of the half-wave supply? How big a capacitor do you need for less than 5% ripple?*

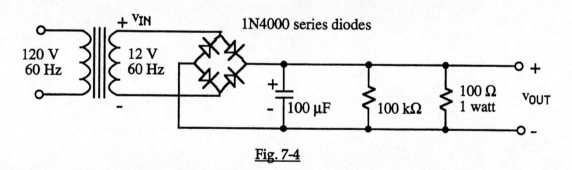

Fig. 7-4

Note: On the bridge rectifier circuit, do NOT try to measure v_{IN} and v_{out} (the ac and dc voltages) simultaneously if your scope has a common ground connection to its two channels. This common ground connection will create a ground loop that will connect one of the diodes directly to the transformer output, leading to its rapid demise.

Voltage Doubler:

4) Voltage-doubler circuits are often used in power supplies of the hundred-kilovolt variety. Construct the low-voltage version shown in Fig. 7-5, and observe the input and output waveforms. (NOT at the same time! See the note above!) Experiment with different load resistors. *Does the circuit behave as expected? Is this a very well regulated supply? Explain the operation of the circuit in your notebook.*

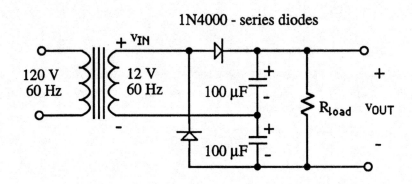

Fig. 7-5

EXPERIMENT 8

THE ZENER-DIODE REGULATOR

4.4.4 Voltage Regulation

Purpose

To study the behavior of the zener diode, and to use it as a voltage regulator to improve the performance of the dc power supply.

Introduction

Even though the diode equation predicts that the diode current is negligibly small for all negative voltages, all real diodes exhibit <u>avalanche breakdown</u> when the reverse voltage becomes large enough. In the breakdown region, the current may rise more rapidly with increasing voltage than it does in forward bias. In normal pn-junction diodes this process may occur at around -100 V, a voltage that is not normally reached in most electronic circuits. However, it is possible to reduce the magnitude of the reverse breakdown voltage, without affecting the forward-bias characteristics, by controlling the amount of dopants introduced into the silicon when the diodes are made. Diodes manufactured in this way are called <u>zener diodes</u>; they find many uses in clipper and limiter circuits. They are superior to ordinary forward-biased diodes in two respects - because they can be tailored to break down over a wide range of voltages, while the forward-biased diode is restricted to the 0.6 V to 0.8 V range, and because the i-v curve is often steeper in reverse breakdown than in forward bias. Commercial zener diodes are available for the range from 3.3 V to several hundred volts.

Although "breakdown" sounds like a catastrophic process, there is no damage to the diode if a means is provided to limit the current to a safe level, as is also necessary in forward bias. It should also be noted that a zener diode in forward bias acts in exactly the same way as a normal pn-junction diode.

Fig. 8-1 shows the i-v characteristic of a typical zener diode. The reverse-breakdown region cannot be described by an analytic expression, but it can be approximated with sufficient accuracy by a straight line with slope = $1/r_z$, where r_z is the small-signal, or incremental resistance in this region, and intercept = $-V_{zk}$. r_z may be as small as 5 ohms, reflecting the steepness of the slope. The intercept $-V_{zk}$ is called the zener knee voltage. The circuit symbol for the zener diode is shown in Fig. 8-2. The voltage polarity is as shown there because the diode is normally operated in the reverse direction, with current flowing downward in accordance with the passive sign convention. To be consistent with this operation, the direction of the voltage axis in Fig. 8-1 should properly be reversed.

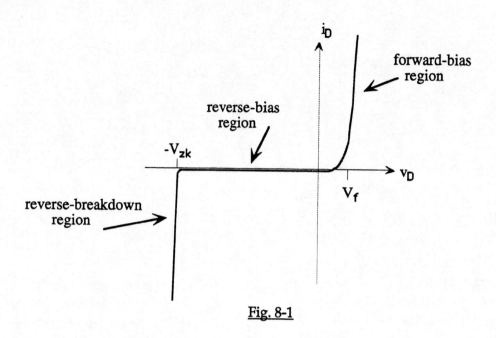

Fig. 8-1

Procedure

I. Measurement of the zener diode i-v characteristic:

1) Set up the circuit shown in Fig. 8-2 below. For several values of v_s in the range 0 to 20 V, make a table of the values of v_z and i_z. Find the value of V_{zk} for your diode. Can you estimate the value of r_z from your data?

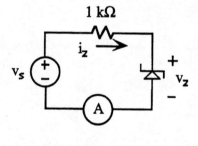

Fig. 8-2

II. The zener regulator:

Modify the full-wave rectifier circuit from Experiment 7 by including a zener diode, as shown in Fig. 8-3.

1) Using the data from Experiment 7 and part I of this experiment, choose an appropriate value of R_1 so that:

a) the zener diode remains in reverse breakdown for $R_L > 1\ k\Omega$, and

b) the power dissipated in the zener diode is less than 200 mW, even with the load resistor removed.

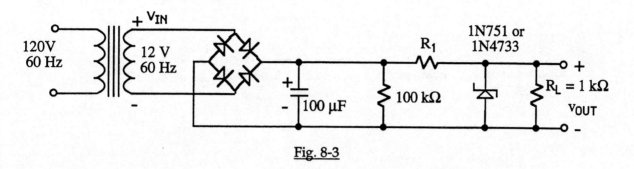

Fig. 8-3

Include in your notebook the calculation that you used to obtain the value of R_1.

2) Measure the voltage across the 100-kΩ resistor and across the load resistor, taking care to record both the dc value and the ac ripple component. *By how much does the zener diode reduce the capacitor ripple voltage?*

3) Measure the output voltage both with and without the load resistor present. *What is the percent regulation of this supply when the zener regulator is in place?* (See equation 7-2.)

4) With the load resistor connected, touch R_L and the zener diode with your finger. Comment on their temperature. Now disconnect R_L and repeat the test. *Can you explain what you observe?*

5) Replace the load resistor with several others in the range of 220 Ω to 10 kΩ, measuring each time the current flowing through the zener diode and the load resistor. Verify that the sum of these two currents always equals the current through R_1, and that the latter is a constant as long as the zener diode remains in reverse breakdown. *Explain why the preceding statement is true.*

EXPERIMENT 9

I-V CHARACTERISTICS OF THE METAL-OXIDE-SILICON FIELD-EFFECT TRANSISTOR

5.2 Field-Effect Transistors
    5.2.1 Physical Structure of the N-Channel Enhancement-Mode MOSFET
    5.2.2 Summary of V-I Equations of the N-Channel Enhancement-Mode MOSFET
    5.2.4 Nonzero Source-to Substrate Voltage (The Body Effect)
    5.4.1 Upward Slope of FET V-I Characteristics

Purpose

To analyze current-voltage relationships of the metal-oxide-silicon field-effect transistor, or MOSFET. The MOSFET is by far the most widely used active element in digital integrated circuits. It is the essential component of practically all random-access memory (RAM) chips used in computers.

Introduction

Three-terminal circuit elements offer a capability not available with the two-terminal ones studied so far; they are active elements, which are able to deliver more power to a load connected to one pair of their terminals than they receive from a control source connected to a second pair. The additional energy is supplied from a separate source connected to the third pair. These active elements may be visualized as valves that require a small amount of energy to open, but which, once open, allow the flow of a large amount of energy. This process underlies many kinds of voltage and current amplifier, which in one form or another are found in almost every electronic system.

Practically all silicon integrated circuits contain either MOSFETs or bipolar junction transistors (BJTs) as the active elements. Although most integrated circuits use only one of the two elements, newer fabrication technology allows the manufacture of chips that contain both. Although they have superficially similar i-v characteristics, they operate by entirely different mechanisms. In the BJT, to be studied in Experiment 11, the essential factor is the forward-biased base-emitter junction, whose current controls a larger collector current. The MOSFET is a purely voltage-controlled element, in which the potential drop across an insulating layer on the silicon surface controls current flow in the silicon. Moreover, the BJT (both the npn and the pnp form) requires the flow of both electrons and holes, whereas the MOSFET requires only one type of carrier.

An n-channel MOSFET, or NMOS, is shown in Fig. 9-1. There is also a p-channel, or PMOS form, in which the doping in each of the three regions of the silicon is reversed. There are therefore structural similarities between the NMOS and npn transistors and between the PMOS and pnp transistors. The insulator is usually a film of silicon dioxide less than 50 nm (500 Å) thick. The MOSFET is actually a four-terminal element, since the substrate also plays a part in its operation.

The substrate of the NMOS is always held at the most negative potential in the circuit (most positive for the PMOS) so that the source-substrate and drain-substrate junctions are never forward biased. Be sure that this is true whenever you build a MOSFET circuit.

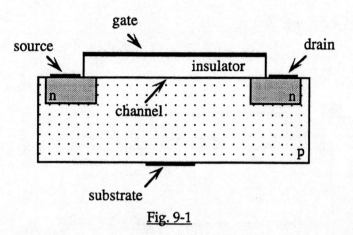

Fig. 9-1

To simplify the description of the operation of an NMOS, consider the source and substrate to be held at ground potential. When the gate potential is zero, and the drain potential is either positive or negative, no current flows through the channel between source and drain because one of the pn junctions is reverse biased. If the gate potential is now made positive, Gauss' Law requires that the positive charge on the top surface of the insulator be balanced by an equal negative charge on the bottom surface, which is in the silicon channel. For low gate potential this is accomplished by forcing positive holes away from the insulator-silicon interface, leaving behind the negatively charged ionized acceptor centers. At higher gate potential, this depletion layer reaches a maximum size; the additional negative charge is supplied by free electrons that enter the channel from the n-type source and drain. An inversion layer is now formed, which acts as an n-type region because the free carriers are electrons. The junctions at the source and drain effectively disappear, and current can now flow from drain to source if the drain is at a positive potential.

The minimum potential difference between the top and bottom of the gate insulator needed to form an inversion layer is called the threshold voltage V_{TR}. Since the source is at the lowest potential, this means that the minimum condition for current to flow is

$$v_{GS} > V_{TR}. \qquad (9\text{-}1)$$

In normal operation, the drain is more positive than the source. At any point along the channel we have, by KVL,

$$v_{GS} = v_{G-channel} + v_{channel-S} \qquad (9\text{-}2)$$

At the drain, which is the most positive point in the channel, this becomes

$$v_{GS} = v_{GD} + v_{DS} \quad \text{or} \quad v_{DS} = v_{GS} - v_{GD} \qquad (9\text{-}3)$$

For fixed v_{GS}, as long as v_D is low enough so that the gate-to-drain potential difference v_{GD} exceeds

V_{TR}, the inversion layer extends across the entire channel, and the drain current is given by

$$i_D = K[2(v_{GS} - V_{TR})v_{DS} - v_{DS}^2], \quad \text{where} \quad v_{DS} < v_{GS} - V_{TR} \qquad (9\text{-}4)$$

This is the triode region of operation (Fig. 9-2).

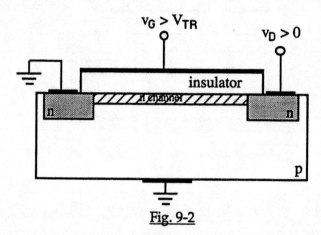

Fig. 9-2

When v_{DS} is increased, with v_G unchanged, v_D eventually becomes large enough so that the the voltage difference $v_G - v_D$ is now less than V_{TR}. The voltage drop across the insulator near the drain is no longer large enough to sustain an inversion layer (Fig. 9-3). There is now a depletion layer at that location,

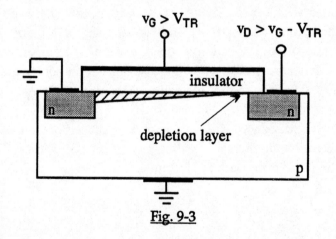

Fig. 9-3

whose built-in field acts in the same way as the collector in the BJT (Experiment 11). Electrons that reach it from the channel are swept into the drain and into the external circuit. The current no longer depends upon v_{DS}; it is now given by

$$i_D = K(v_{GS} - V_{TR})^2, \quad \text{where} \quad v_{DS} > v_{GS} - V_{TR} \qquad (9\text{-}5)$$

This is the constant-current region, sometimes called the saturation region. The latter term can be confusing because it misleadingly suggests a parallel with the saturation region of the BJT. The transition between the constant-current region and the triode region occurs where

$$v_{DS} = v_{GS} - V_{TR} \tag{9-6}$$

Figure 9-4 shows the i-v characteristic of an NMOS. As in the case of the BJT, the PMOS characteristics have the same shape, but lie in the third quadrant, with all signs reversed.

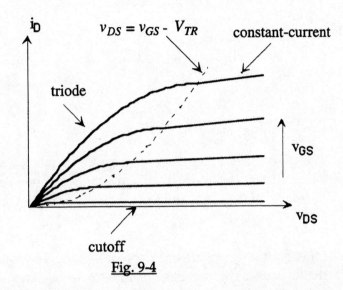

Fig. 9-4

The <u>cutoff</u> region corresponds to the condition $v_{GS} \leq V_{TR}$. The locus of the boundary points between the triode and constant-current regions is shown by the dashed curve. As in the case of the BJT, there is a slight dependence of i_D on v_{DS} in the constant-current region; the lines converge to a point called the Early voltage. Each curve corresponds to a different value of v_{GS}. Note that they are not equally spaced for equal increments of the controlling variable, as they are in the BJT, because of the square-law relationship in (9-5).

When the source-to-substrate junction is reverse biased, the depletion layer between the channel and the substrate becomes wider, causing the channel to become "pinched off" from the bottom, reducing its conductance. A larger v_{GS} is needed to return the current to its zero-bias value. The reverse bias has the effect of increasing the effective threshold voltage of the device. This <u>body effect</u> is described by equation (9-7).

$$V_{TR} = V_{TR0} \pm \gamma \left[\sqrt{|v_{SB}| + |2\phi_F|} - \sqrt{|2\phi_F|} \right] \tag{9-7}$$

The parameters γ and ϕ_F are determined by the structure and dopant concentrations of the MOSFET. V_{TR0} is the nominal threshold voltage in the absence of the body effect, and v_{SB} is the source-to-body (substrate) potential. The positive sign applies to the NMOS and the negative sign to the PMOS.

A major difference between the BJT and the MOSFET is that the "threshold" for turning on the BJT is fixed by the properties of the pn junction at about 0.6 V, whereas the threshold voltage V_{TR} of the MOSFET can be controlled by introducing dopants into the channel. In fact, it can be made positive or negative for both kinds of MOSFET. This leads to four different kinds of

MOSFET, as shown in Fig. 9-5.

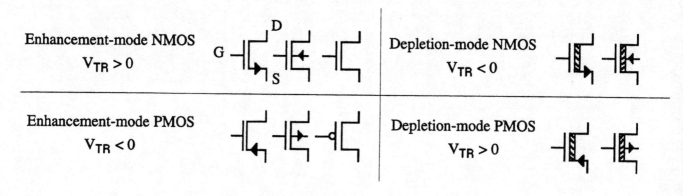

Fig. 9-5

The commonly used circuit symbols are shown for each type. The arrows are placed either at the source or at the substrate. When at the source, they show the direction of current flow; when at the substrate, they point from the p region to the n region, as they do in the pn-junction diode symbol.

The i-v curves for enhancement-mode and depletion-mode devices are indistinguishable. The only difference is that the curve that corresponds to $v_{GS} = 0$ coincides with the x axis (cutoff) in the enhancement-mode case, but is above the x axis ($i_D \neq 0$) in the depletion-mode case. The characteristic for the PMOS has the same shape as for the NMOS, except that the curves lie in the third quadrant, where the signs of all voltages and currents are reversed.

The MOSFET is a voltage-controlled element, as opposed to the BJT, which is current-controlled. No significant gate current flows at the MOSFET gate terminal because it is connected to an insulator. Another difference between the two devices is that the MOSFET is truly symmetrical, as suggested by several of the circuit symbols in Fig. 9-5, whereas the BJT is not. The source and drain may be freely interchanged, providing that the substrate is not permanently connected to one of them.

Procedure

The CD4007 integrated-circuit chip is a useful vehicle for this experiment because it contains both NMOS and PMOS transistors (although only enhancement-mode versions). The chip is mounted in a 14-pin DIP package with the terminals of six transistors accessible at the pins as shown in Fig. 9-6. The PMOS devices are in the top row of the diagram, and the NMOS in the bottom row.

Note that the substrate, or body, for each PMOS is to be connected to the most positive point in your circuit, and the substrate for each NMOS is to be connected to the most negative point. If you fail to do this, one of the pn junctions will be forward biased, and probably destroyed. The source is connected internally to the substrate in the two leftmost devices. Therefore, for measurement of the body effect you must use one of the other devices so that you may control the potential difference v_{SB} between the source and the body.

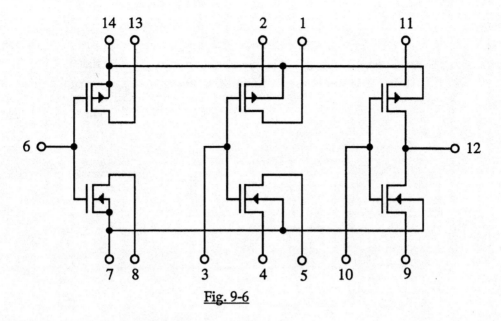

Fig. 9-6

I. Measurement of K, V_{TR}, and γ

BEFORE COMING TO THE LABORATORY:

Construct the circuit shown in Fig. 9-7, using one of the NMOS devices on the CD4007 chip that does not have an internal connection between source and body. If you have never used a multi-pin integrated circuit (IC) before, consult Appendix C, which gives the pin connections to the "mini-dip" package of the LM741 op amp. Be sure to double check your wiring, especially to the IC, before turning on the power in the lab.

Since the NMOS has its gate and drain connnected together, it always operates in the constant-current region, because $v_{DS} > v_{GS} - V_{TR}$ when $v_{GS} = v_{DS}$ and $V_{TR} > 0$ (enhancement-mode device). The potentiometer connected to a negative voltage allows the body potential to be set anywhere between zero and -15 V. The feedback circuit effectively clamps the source potential at ground.

IN THE LABORATORY

1) Set the body potential to zero and increase V_{DD} until a measurable drain current is reached. Measure i_D for several values of $v_{GS} = v_{DS}$ up to about 10 V.

2) Repeat the sequence of of i_D vs v_{GS} measurements for different values of v_{SB} up to about 10 V. Be sure that the substrate is never positive with respect to the source.

II. Measurement of the i-v characteristic

BEFORE COMING TO THE LABORATORY

In this part you will measure the characteristic curves of i_D vs v_{DS} for different v_{GS}. Build the circuit of Fig. 9-8, using either the NMOS transistor on the CD4007 chip that has the internal connection between source and substrate, or one of the others with an external connection between these two terminals.

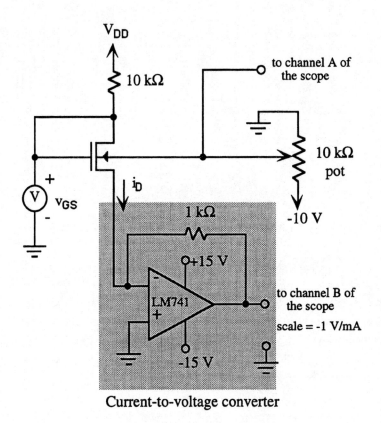

Current-to-voltage converter

Fig. 9-7

IN THE LABORATORY

1) Before connecting the signal generator to the circuit, observe its output directly on the scope. Using the dc offset capability of the generator, set the output so that it is about equal to a 15 V p-p waveform, offset by a dc voltage of about +7.5 V, so that the output range is between 0 and +15V.

2) Now connect the signal generator to the circuit. Note that the signal generator functions as a varying supply voltage V_{DD}, which allows the load line imposed on the transistor to sweep across values of open-circuit voltage between 0 and 15 volts, thereby continually sweeping across various operating points of the transistor.

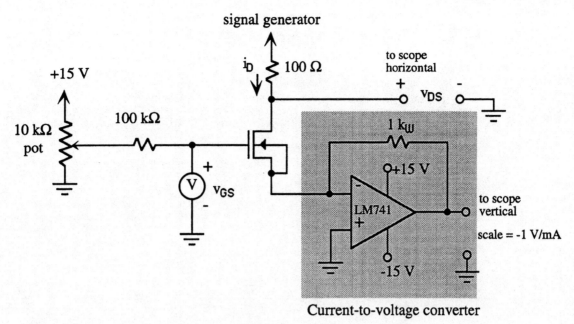

Fig. 9-8

3) Set the oscilloscope to the X-Y mode, with each channel set to 1V/div. Set the vertical channel on "invert", so that the negative voltage output of the op amp circuit will register as a positive deflection of the scope. Connect the horizontal channel between the drain lead of the NMOS and real ground (NOT the op amp input lead), and connect the vertical inverted channel of the scope between the op amp output lead and ground. In this way the scope will display the transistor output-port current vs output-port voltage. Gradually increase the setting of the potentiometer, recording the value of v_G as measured on the voltmeter. Record in your notebook the resulting transistor i-v curve for several values of v_{GS} (note that $v_S \approx 0$ because of the negative feedback).

4) Repeat steps 1 through 3 using one of the PMOS devices with its source and substrate connected together. The circuit should be modified as shown in Fig. 9-9 so that the signs of all the terminal voltages are reversed. In this case, set the output of the voltage generator connected the drain so that it is about equal to a 15 V p-p waveform, offset by a dc voltage of about -7.5 V, so that the output range is between 0 and -15V. Just as in the NMOS case, the source is held at virtual ground by the op amp feedback circuit. The potentiometer allows the gate voltage to be set negative with respect to the source. The vertical channel of the oscilloscope should again be inverted so that the i-v curves will be displayed in the third quadrant, where v_{DS}, v_{GS}, and i_D are negative.

Identify the transistors that you used for these measurements so that you can use them later if necessary.

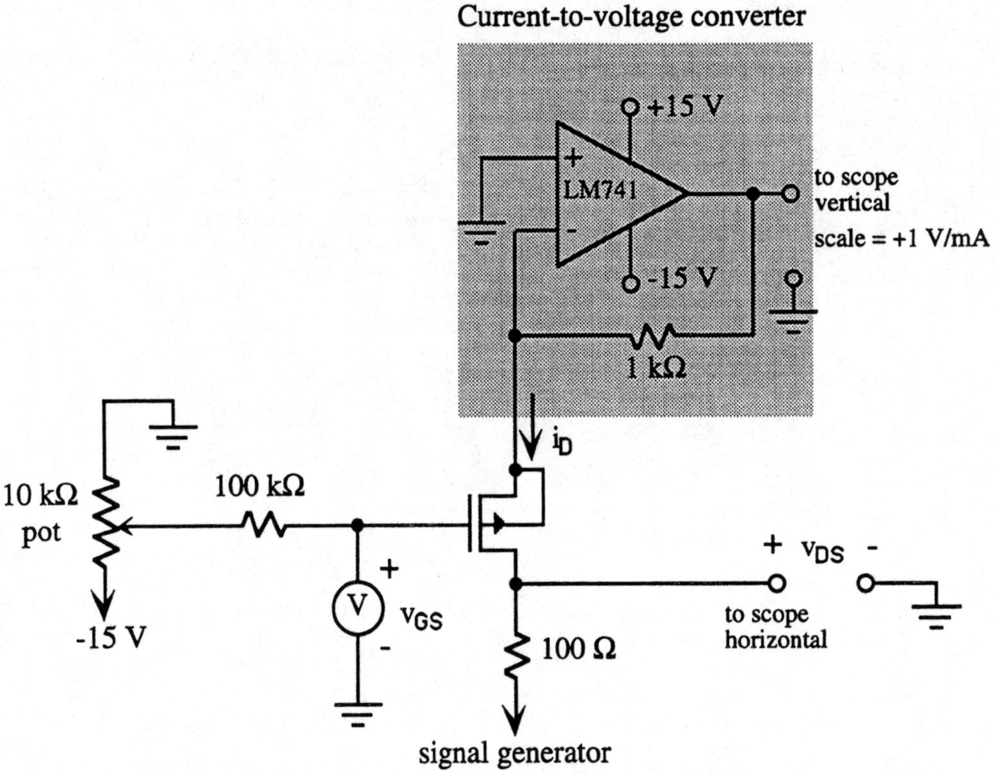

Current-to-voltage converter

Fig. 9-9

Analysis of the Data

A) Using the data from Part I for the NMOS, obtain values for K and V_{TR} from a plot of the square root of i_D vs v_{GS} for each value of v_{SB}. Rearranging equation (9-5), we get

$$\sqrt{i_D} = \sqrt{K}v_{GS} - \sqrt{K}V_{TR} \tag{9-8}$$

Above the threshold voltage, the plots should be straight lines with slope = $K^{0.5}$. The intercept at $i_D = 0$ is V_{TR}.

B) Calculate a value for the NMOS body coefficient γ from (9-7), using the calculated values of V_{TR} for different v_{SB}, and 0.3 V as an appropriate value for ϕ_F.

C) For both types of transistor, estimate the value of v_{DS} where the triode and constant-current regions meet. Is it consistent with equation (9-6) for the NMOS case when you use your calculated value for V_{TR}?

D) For both types of transistor, try to calculate the slopes of the i-v curves in the constant-current region for several dfferent values of v_{GS}. The inverse of the slope is often designated r_0, the incremental resistance in this region. The magnitude of the slope should increase with increasing $|v_{GS}|$. If the slopes can be determined with reasonable accuracy, you should find that all the i-v curves in the forward-active region converge to approximately the same value of v_{DS} at $i_C = 0$. This point is called the Early voltage V_A. Try to find its value for your transistors.

EXPERIMENT 10

I-V CHARACTERISTICS OF THE JUNCTION FIELD-EFFECT TRANSISTOR

5.2.6 Junction Field-Effect Transistor

Purpose

The junction field-effect transistor (JFET) is another useful 3-terminal element. You will study its characteristics using methods similar to those used for the MOSFET.

Introduction

The JFET is in some ways a hybrid of the BJT and the MOSFET. It resembles the BJT in that it contains only pn junctions, and no insulating layers. Yet its electrical characteristics are similar to those of the MOSFET; current flow between two of the terminals is controlled by a voltage applied to the third terminal, into which negligible current flows.

The concept of the JFET can be explained with Fig. 10-1. Although this cylindrical form is not used in a real JFET, it is useful for explaining the conduction mechanism. The uniform, moderately doped cylinder of n-type silicon has ohmic source and drain contacts at the ends. A band of heavily doped p-type material surrounds the cylinder, but does not extend all the way to the center, as shown in the end view.

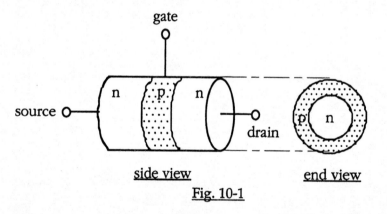

Fig. 10-1

Begin with the source and gate at ground potential, and a small positive potential on the drain. Current flows from the drain to the source through the uniformly wide n-type channel. Keeping the drain voltage constant at this low value, and biasing the gate negatively, increases the width of the depletion layer around the junction, thus decreasing the diameter, and therefore the conductance, of the channel. See the cross-sectional view in Fig. 10-2. (The depletion layer extends into the p-type gate region also, but not very far because of the heavy doping there.) The JFET therefore acts as a voltage-controlled variable resistance, just as the MOSFET does in the triode region. When the gate potential becomes so negative that the depletion layer extends all the way to the center of the channel, pinchoff is reached, and the current goes to zero. This is analogous to the cutoff region of the MOSFET.

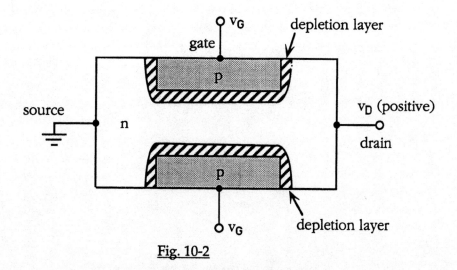

Fig. 10-2

If the gate potential is held constant (at zero or some negative value), and v_D is made more positive, the drain end of the junction is more strongly reverse-biased than the source end. Therefore the depletion layer is wider near the drain. This situation is shown in Fig. 10-3. At some value of v_D, the channel will become pinched off near the drain, and there will no longer be a continuous n-type region from source to drain. Electrons from the source that reach the pinched off region are accelerated toward the drain by the same kind of built-in field that exists in the MOSFET when its inversion layer disappears near the drain. This is the <u>constant-current region</u>, where the current is no longer a function of v_{DS}.

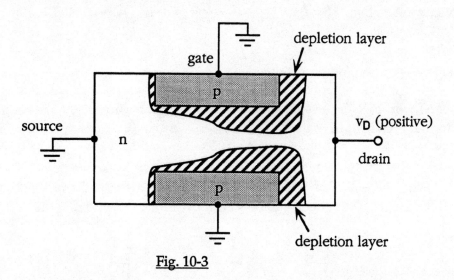

Fig. 10-3

The JFET is clearly a field-effect device, as is the MOSFET. In the JFET the field exists in the depletion layer, whereas in the MOSFET it exists in the gate insulator. The JFET gate current is the reverse-bias leakage current of a pn junction. Although this current is much greater than the leakage current of the MOSFET's gate insulator, it is still small enough to be ignored in most cases. The i-v characteristic of the JFET is shown in Fig. 10-4. The non-zero slope of the curves in the constant-current region is caused by the same channel-length modulation effect that occurs in the

MOSFET. So the JFET is similar to a depletion-mode MOSFET, but with one significant difference - in the MOSFET, v_{GS} may be positive or negative, thus increasing or decreasing the drain current, while in the JFET the current can only be decreased; the gate-source junction must never be forward biased.

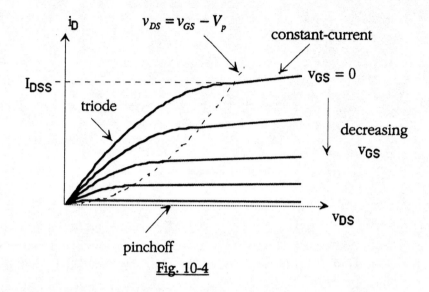

Fig. 10-4

The i-v equations for the JFET are essentially the same as those for the MOSFET, although the terminology is somewhat different. The equations are commonly written in the following form:

constant-current region: $i_D = I_{DSS}\left(1 - \dfrac{v_{GS}}{V_p}\right) = \dfrac{I_{DSS}}{V_p^2}(V_p - v_{GS})^2$, $v_{GS} \leq V_p$ and $v_{DS} > v_{GS} - V_p$ (10-1)

triode region: $i_D = \dfrac{I_{DSS}}{V_p^2}\left(2(v_{GS} - V_p)v_{DS} - v_{DS}^2\right)$, $v_{GS} \leq V_p$ and $v_{DS} < v_{GS} - V_p$. (10-2)

The pinchoff voltage V_p is the equivalent of the threshold voltage V_{TR}. The maximum current I_{DSS} is used for the JFET instead of the K factor.

Remember that in this n-channel JFET, both v_{GS} and V_p are negative. There is also a p-channel version of the JFET, in which the n and p regions are interchanged and the signs of all the voltages are changed. The circuit symbols for the JFET are shown in Fig. 10-5.

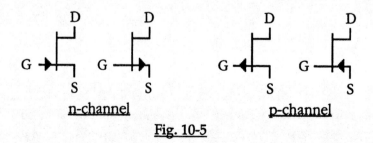

Fig. 10-5

As usual, the arrow points from the p region toward the n region, or in the direction of current flow.

Procedure

BEFORE COMING TO THE LABORATORY

Build the circuit shown in Fig. 10-6, to be used to measure the i-v characteristic of an n-channel JFET. Use a 2N3819 or equivalent JFET. This circuit is essentially the same as the ones used to measure the characteristics of the MOSFET. The main difference is that the gate voltage must never become positive, to prevent forward-biasing the gate-source junction.

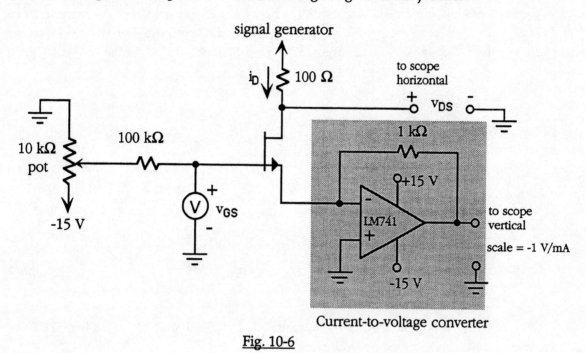

Fig. 10-6

IN THE LABORATORY

1) Before connecting the signal generator to the circuit, observe its output directly on the scope. Using the dc offset capability of the generator, set the output so that it is about equal to a 15 V p-p waveform (sinewave or triangular wave), offset by a dc voltage of about +7.5 V, so that the output range is between 0 and +15V.

2) Now connect the signal generator to the circuit. Note that the signal generator functions as a varying supply voltage V_{DD}, which allows the load line imposed on the transistor to sweep across values of open-circuit voltage between 0 and 15 volts, thereby continually sweeping across various operating points of the transistor.

3) Be sure that the potentiometer is connected to a negative voltage as shown, so that the gate voltage will not become positive.

4) Set the oscilloscope to the X-Y mode, with each channel set to 1V/div. Set the vertical channel on "invert", so that the negative voltage output of the op amp circuit will register as a positive deflection of the scope. Connect the horizontal channel between the drain lead of the JFET

and real ground (NOT the op amp input lead), and connect the vertical inverted channel of the scope between the op amp output lead and ground. In this way the scope will display the transistor output-port current vs output-port voltage. Beginning with $v_G = 0$, gradually increase the setting of the potentiometer, recording the value of v_G as measured on the voltmeter. Record in your notebook the resulting transistor i-v curve for several values of v_{GS} (note that $v_S \approx 0$ because of the negative feedback).

5) Replace the signal generator with a dc voltage source, set to a voltage that places the JFET in the constant-current region of operation when $v_{GS} = 0$. Record the drain current for zero gate voltage, then for successively more negative values values until the current is too low to measure accurately. Repeat these measurements for another value of v_D in the constant-current region.

Analysis of the Data

A) Using the i-v curves that you obtained in part 4, find or estimate values for I_{DSS} and V_p.

B) Obtain more precise values for I_{DSS} and V_p from plots of $\sqrt{i_D}$ vs v_{DS}, using the data of part 5.

C) Estimate the value of r_0 or V_A, the Early voltage, from the slope of the i_D vs v_{DS} curves, as you did for the BJT.

D) List what you consider to be the advantages and disadvantages of the MOSFET and JFET for use as the active element in amplifiers.

EXPERIMENT 11

I-V CHARACTERISTICS OF THE BIPOLAR JUNCTION TRANSISTOR

5.1 Definition of Three-Terminal Devices
   5.3 Bipolar-Junction Transistor
      5.3.1 Physical Structure of the Bipolar Transistor
      5.3.2 NPN BJT V-I Characteristics
      5.4.2 Upward Slope of BJT V-I Characteristics

Purpose

To study the current-voltage relationships of one of the most important active elements, the bipolar junction transistor (BJT). The BJT is widely used in integrated circuit chips for linear and high-speed digital applications.

Introduction

The bipolar junction transistor (BJT) is a three-terminal device that consists of three differently doped regions in a single piece of semiconductor, separated by pn junctions. The most commonly used configuration, the npn transistor, has an n-type <u>emitter</u> region, followed by a p-type <u>base</u>, and an n-type <u>collector</u>. The pnp transistor has the same sequence of regions, but as the name implies, the opposite order of doping. Fig. 11-1 shows the basic structure and circuit symbol for each.

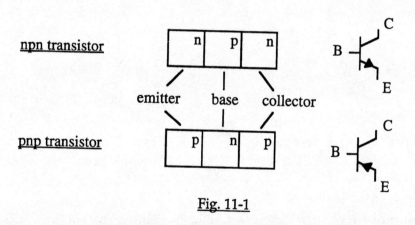

Fig. 11-1

The arrow on the circuit symbol identifies the emitter terminal in both cases. It points from the p region toward the n region, just as it does in the symbol for the pn-junction diode.

The npn BJT (Fig. 11-2) is constructed so that when the base-emitter junction is forward biased, most of the current is carried by the electrons rather than the holes. This is accomplished by making the concentration of donor impurities in the emitter about 1000 times greater than the concentration of acceptor impurities in the base. The base region is made very narrow, typically a few tenths of a micrometer, so that most of the electrons that enter it from the emitter reach the base-collector junction before they disappear by recombining with the holes in the base or by leaving the silicon through the base terminal. The built-in electric field at the base-collector junction accelerates the electrons into the collector, where they leave through the collector terminal. These electrons constitute the collector current. The base current, consisting of the small number of holes needed to sustain the flow across the base-emitter junction, is much smaller than the collector current.

Although Fig. 11-2 shows the base and collector terminals shorted to each other, there is still a built-in electric field at the base-collector junction that will sweep the electrons out of the silicon at the collector. It makes little difference whether this junction is at zero potential or reverse biased. The collector current is determined not by the collector-base voltage drop, but by the base-emitter voltage drop, which controls the number of electrons injected into the base.

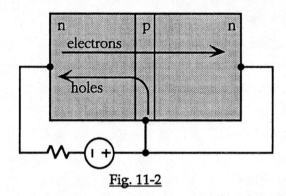

Fig. 11-2

There are four modes, or regions, of operation for the BJT, determined by the state of the two pn junctions, as shown in Table 11-1. Since this table applies to both the npn and pnp transistors, it follows that the signs of all the voltages are opposite for the two types.

Table 11-1

OPERATING MODE	B-E JUNCTION	B-C JUNCTION
CUTOFF	reverse bias	reverse bias[1]
FORWARD-ACTIVE	forward bias	reverse bias [2]
REVERSE-ACTIVE	reverse bias	forward bias
SATURATION	forward bias	forward bias

[1] The state of the B-C junction is practically irrelevant here. No emitter current flows unless the B-E junction is forward-biased.

[2] The B-C junction may have zero bias in this case, or may even be slightly forward biased by a few tenths of a volt, as long as the base current is negligible.

Fig. 11-3 shows a typical set of i-v characteristics for the npn bipolar transistor. The characteristic for the pnp has the same shape, but lies in the third quadrant, where the signs for all voltages and currents are reversed.

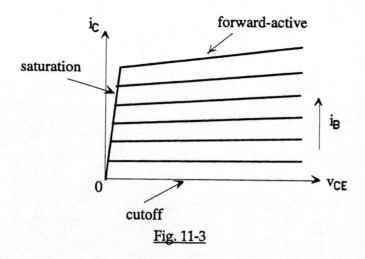

Fig. 11-3

The forward-active region is the one normally used for linear circuit applications. In this region, the current flowing into the collector is proportional to the much smaller current flowing into the base. The base current, in turn, is related to the base-emitter voltage difference v_{BE} by the usual exponential equation for the forward-biased diode (eqn. (4-1)). The currents in this region are related by $i_C = \beta_F i_B$, where β_F is in the approximate range 30-200, depending upon the type of transistor. The collector current i_C depends slightly on v_{CE} in the forward-active region, although this dependence is usually ignored in the "ideal" BJT. In analogy with the MOSFET, the lines in the forward-active region converge to a point called the Early voltage. There is actually a continuous family of curves, one for each value of i_B, which may, of course, vary continuously. In the saturation region, where v_{CE} is less than about 0.3 V, i_C is practically independent of i_B. In cutoff, where i_B is zero or extremely small, i_C is practically zero.

Although the BJT appears to be a symmetric device, the dopant concentrations in the emitter and collector are quite different, so that β_R, the current ratio in the reverse-active region, is only of the order of unity or less. This operating region finds no uses in linear amplifiers, but it does serve a purpose in transistor-transistor logic (TTL) circuits.

The BJT may be considered a current-controlled circuit element, in which a small base current controls a much larger collector current.

Procedure

BEFORE COMING TO THE LABORATORY

Construct the circuit shown in Fig. 11-4, using a 2N2222 npn transistor or a similar type. If you have never used a multi-pin integrated circuit (IC) before, consult Appendix C, which gives the pin connections to the "mini-dip" package of the LM741 op amp. Be sure to double check your wiring, especially to the IC, before turning on the power in the lab. If you have made incorrect connections you may destroy one or more components. The shaded portion of the circuit is the same current-to-voltage converter that was used in Experiment 9. Review the description of its function presented in that experiment.

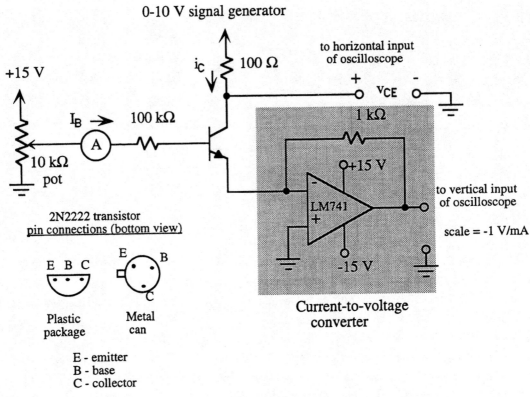

Fig. 11-4

The purpose of the op amp current-to-voltage converter is to provide a voltage signal to the oscilloscope that is directly proportional to the current flowing through the transistor. As designed above, the op amp circuit will deliver -1 volt at its output terminals per mA of emitter current. Moreover, the negative feedback in the op amp circuit will maintain the emitter terminal at essentially ground potential. This insures that the voltage measured at the collector with respect to ground virtually equals v_{CE}, and that the voltage at the base virtually equals v_{BE}.

IN THE LABORATORY

1) Before connecting the signal generator to the circuit, observe its output directly on the scope. Using the dc offset capability of the generator, set the output so that it is about equal to a 10-V p-p waveform, offset by a dc voltage of about +5 V, so that the output range is between 0 and +10V.

2) Now connect the signal generator to the circuit. Note that the signal generator functions as a varying supply voltage V_{CC}, which allows the load line imposed on the transistor to sweep across values of open-circuit voltage between 0 and 10 volts, thereby continually sweeping across various operating points of the transistor.

Set the oscilloscope to the X-Y mode, with each channel set to 1V/div. Set the vertical channel on "invert", so that the negative voltage output of the op amp circuit will register as a positive deflection of the scope. Connect the horizontal channel between the collector lead of the BJT and real ground (NOT the op amp input lead), and connect the vertical inverted channel of the scope between the op amp output lead and ground. In this way the scope will display the transistor output-port current vs output-port voltage.

Gradually increase the setting of the potentiometer, recording the value of I_B as measured on the ammeter. Record in your notebook the resulting transistor i-v curve for several values of I_B.

Analysis of the Data

A) Calculate β_F for your transistor by finding the ratio I_C/I_B at several values of I_B. All the collector currents used in this calculation should be measured at the same value of V_{CE}.

B) Estimate V_{CEsat}, the point where the saturation and forward-active regions meet.

C) Try to calculate the slopes of the i-v curves in the forward-active region for several different values of I_B. The inverse of the slope is often designated r_0, the incremental resistance in this region. The slope should decrease with increasing I_B. If the slopes can be determined with reasonable accuracy, you should find that all the i-v curves in the forward-active region converge to approximately the same value of v_{CE} at $i_C = 0$. This point is called the Early voltage V_A. Try to find its value for your transistor.

D) Label the transistor whose parameters you have measured, and save it for use in Experiment 12.

EXPERIMENT 12

TRANSISTOR INVERTERS

6.1 Inverter Configuration
 6.1.1 BJT Inverter
 6.1.2 MOSFET Inverter

Purpose

The voltage transfer functions v_{OUT}/v_{IN} of BJT and MOSFET inverters will be examined. The inverter is the basic element of most linear voltage amplifiers, which are designed so that a change in the input-port voltage causes a proportional, and usually larger, change in the output-port voltage. The inverter is also the heart of most logic gates, which are designed to operate at the extremes of the voltage-transfer function, where the output voltage changes from a low value (logic state "0") to a high value (logic state "1") when the input voltage changes from high to low. The intermediate region, where the output is proportional to the input, is avoided in logic gates, and should be as sharp a transition as possible. In essence, a logic gate is a high-gain (high $\Delta v_{OUT} / \Delta v_{IN}$) voltage amplifier.

Introduction

I. The BJT Inverter

The voltage-transfer function of a simple BJT inverter will be measured with the circuit of Fig. 12-1.

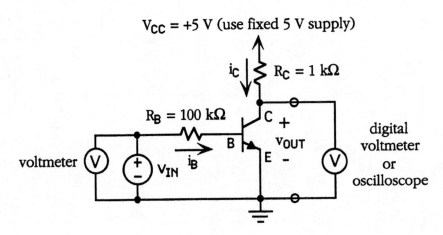

Fig. 12-1

Its voltage-transfer function is derived under the assumption that the transistor operates in the forward-active region, where

$$i_C = \beta_F i_B \qquad (12\text{-}1)$$

Obtaining i_C and i_B from the circuit, we get

$$i_C = \frac{V_{CC} - v_{OUT}}{R_C} = \beta_F \frac{v_{IN} - V_f}{R_B} \tag{12-2}$$

or

$$v_{OUT} = V_{CC} - \beta_F \frac{R_C}{R_B}(v_{IN} - V_f), \tag{12-3}$$

where V_f is the "turn-on" voltage of the base-emitter diode (≈ 0.7 V). This is a linear relationship that applies when $v_{IN} > V_f$. For smaller v_{IN} the transistor is cut off, i_C is virtually zero, and $v_{OUT} = V_{CC}$. If v_{IN} becomes too large, the transistor reaches saturation, where (12-1) no longer applies, and v_{OUT} will change very little with changes in v_{IN}. The voltage transfer function will look roughly like Fig. 12-2.

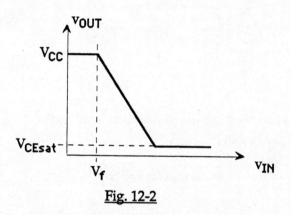

Fig. 12-2

The slope in the linear region, $\Delta v_{OUT} / \Delta v_{IN}$, is $-\beta_F \frac{R_C}{R_B}$, which can be considerably greater than unity if the resistor values are chosen appropriately. This slope is called the voltage gain.

If this simple circuit is to be a suitable linear amplifier, the voltage gain should be constant for all values of v_{IN}. Clearly the transistor must not be allowed to reach either cutoff or saturation. Therefore, the allowable range of v_{IN} values is limited. You will determine this range from your measurements. The concept of voltage gain will be studied more closely in a later experiment.

The input signal to an amplifier is usually a time-varying waveform. To insure that the output waveform is a faithful image of the input, the amplifier must be biased, or preset with a dc input voltage so that the preset output voltage lies between V_{CC} and V_{CEsat} in Fig. 12-2. Subsequent changes in the output voltage will then be proportional to subsequent changes in the input voltage. Biasing can be accomplished by a slight change in the circuit, shown in Fig. 12-3.

V_{BB}, shown as a battery, may be any dc source; its purpose is to set the bias level, or operating point of the inverter on the linear part of the transfer curve. The signal to be amplified is $v_s(t)$. The output voltage will be a linear superposition of the dc operating-point voltage V_{OUT} and the time-varying incremental output signal v_{out}.

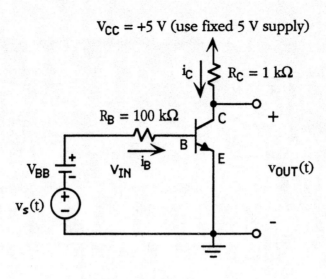

Fig. 12-3

II. The MOSFET Inverter

When a MOSFET is used instead if a BJT in the circuit of Fig. 12-1, with no other changes, other than in terminology, we obtain Fig. 12-4.

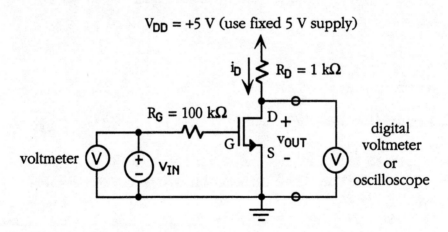

Fig. 12-4

Since there is no gate current, the drain current is controlled by v_{GS}, which is the same as v_{IN}. When the MOSFET is in the constant-current region, the voltage-transfer relationship is

$$i_D = \frac{V_{DD} - v_{OUT}}{R_D} = K(v_{IN} - V_{TR})^2 \tag{12-4}$$

or

$$v_{OUT} = V_{DD} - KR_D(v_{IN} - V_{TR})^2 \tag{12-5}$$

60

Equation (12-5) holds for $v_{IN} > V_{TR}$, and v_{IN} less than the value needed to force the transistor into the triode region. In contrast to (12-3), this is not a linear relationship. Still, it is worthy of study. You will learn later how the MOSFET inverter can function as a linear amplifier. Just as for the BJT, the MOSFET must be biased to operate near the middle of the constant-current region. The MOSFET counterpart of Fig. 12-3 accomplishes this goal.

Procedure

BEFORE COMING TO THE LABORATORY

Construct the circuit of Fig. 12-1, using the BJT whose parameters you measured in Experiment 11. Be sure to use the proper pin connections to the transistor. These were shown in Fig. 11-4. v_{IN} should be a variable dc source. Sketch the *expected* voltage transfer curve, and label as many voltage values as you can predict from your measurements in Experiment 11.

IN THE LABORATORY

1) Beginning with $v_{IN} = 0$, measure v_{OUT} at appropriate increments of v_{IN} until v_{OUT} becomes insensitive to further increase in v_{IN}. Use a digital voltmeter ($R_{in} > 10$ MΩ) to measure v_{OUT}, so that the internal resistance of the meter will not load down the output node. Record v_{OUT} vs v_{IN}.

2) Change R_B to about 10 kΩ and repeat the measurements of part 1, again recording the results.

3) Simplify the procedure by using the oscilloscope in x-y mode to display the voltage transfer function. Connect the x input to v_{IN} and the y input to v_{OUT}. As you vary v_{IN}, the transfer function will be traced on the scope. If you wish, you can automate the procedure by substituting a signal generator for the dc source used for v_{IN}. *Does it make any difference whether you use a sinusoidal wave or a triangular wave for the input? How about a square wave?*

4) Now change your circuit to the arrangement of Fig. 12-3 with $R_B = 100$ kΩ. Be sure to "float" the output terminals of the dc source V_{BB} by disconnecting the "common" terminal from the "ground" terminal. This will avoid a ground loop. Display both v_s and v_{OUT} vs time simultaneously on the oscilloscope, using its dual-trace capability. Also, be sure to use dc coupling for both channels, so you can measure both the dc and the time-dependent components of the signals. With $v_s = 0$, choose a value for V_{BB} that places the bias value V_{OUT} about midway between 0 and 5 V. Then increase v_s and record your observations. Experiment with other combinations of V_{BB} and v_s. Note the effects on v_{OUT}, and record your observations.

5) Repeat parts 3) and 4) of this experiment, substituting for the BJT the enhancement-mode NMOS device on the CD4007 chip that you measured in part II of Experiment 9. The NMOS should have its source and substrate connected together. Use the circuit of Fig. 12-4 and the MOSFET equivalent of Fig. 12-3. Record the voltage-transfer characteristics and identify the cutoff, constant-current, and triode regions of operation on the graphs. You should be able to obtain a

good estimate of V_{TR} from these measurements.

Analysis of the Data

A) From the measurements of the BJT inverter, estimate a value for V_f of the base-emitter diode. Note that the exact value is not critical in equation (12-3); a change from 0.6 V to 0.7 V causes a difference of only about 2 percent in v_{OUT}.

B) Determine from your experimental plots the voltage gain of the BJT inverter for the two cases of $R_B = 100$ kΩ and $R_B = 10$ kΩ. Compare them with the values calculated from (12-3). Which of the two versions would you use as a high-gain voltage amplifier? Which would make the better logic gate?

C) Do your experimental voltage-transfer curves differ in any significant ways from Fig. 12-2? In particular, does the linear region have a constant slope over its entire length? Discuss any major differences, citing reasons for them if you can.

D) Estimate a value for V_{TR} from the measurements of the MOSFET inverter.

EXPERIMENT 13

TRANSISTOR BIASING

7.3 Biasing
  7.3.1 General Biasing Concepts
  7.3.2 Biasing Techniques for the BJT
  7.3.3 Biasing Techniques for MOSFETs and JFETs

Purpose

Although the concept of biasing the transistor inverter at an appropriate operating point was introduced in Experiment 12, the elementary scheme used there, while useful as an introduction to the idea, is inadequate for achieving stable, reproducible amplifiers. In this experiment you will discover the limitation of the elementary method, and will investigate a practical method for obtaining a stable operating point.

Introduction

BJT Biasing

The reason for the inadequacy of the earlier biasing scheme lies in equation (12-3), which is repeated here. Recall that the equation applies to the circuit in Fig. 12-1.

$$v_{OUT} = V_{CC} - \beta_F \frac{R_C}{R_B}(v_{IN} - V_f) \tag{13-1}$$

At the operating point, v_{IN} and v_{OUT} are the dc quantities V_{BB} and V_{OUT}. For a given V_{BB}, V_{OUT} should be independent of the transistor parameters. Unfortunately, β_F is a poorly controlled parameter. When a transistor fails and is replaced by a nominally identical one, β_F is likely to be different. For the 2N2222, for example, the manufacturer specifies that β_F may be anywhere in the range of 50 to 150. Here is what can happen when the circuit in Fig. 12-3 is built with $V_{CC} = 5$ V, $R_C = 1$ kΩ, $R_B = 10$ kΩ. Using a transistor with $\beta_F = 50$ and setting $V_{BB} = 1.2$ V places V_{OUT} at 2.5 V, which is a desirable operating point, midway between 0 and V_{CC}. If the transistor is replaced by a nominally identical one that happens to have $\beta_F = 150$, and V_{BB} is unchanged, the new calculated value of V_{OUT} is -2.5 V. The negative value means that the new transistor is in the saturation region, where (13-1) does not apply. Obviously we need a biasing scheme that will allow for the change in β_F.

Insensitivity of the operating point to changes in β_F can be achieved by adding a resistor R_E into the emitter branch of the circuit in Fig. 12-3, as shown below in Fig. 13-1. Since we are considering only the operating point, we set $v_s = 0$.

To see the effect of R_E, apply KVL to the base-emitter loop:

$$-V_{BB} + I_B R_B + V_f + I_E R_E = 0 \tag{13-2}$$

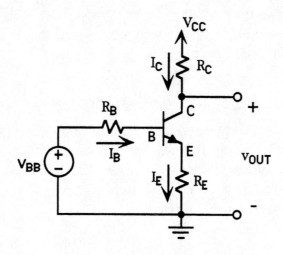

Fig. 13-1

Using

$$I_E = (\beta_F + 1)I_B \text{ and } I_C = \beta_F I_B \tag{13-3}$$

leads to

$$I_C = \beta_F I_B = \frac{\beta_F(V_{BB} - V_f)}{R_B + (\beta_F + 1)R_E} \tag{13-4}$$

If we choose $R_B \ll (\beta_F + 1)R_E$, \hfill (13-5)

then

$$I_C \approx \frac{V_{BB} - V_f}{R_B}, \tag{13-6}$$

because, for $\beta_F \gg 1$, $\beta_F + 1$ can be replaced by β_F. We now have a collector current that is independent of β_F. Compare (13-6) with (12-2), which shows that I_C is directly proportional to β_F when there is no emitter resistor.

The circuit of Fig. 13-1 is still not entirely satisfactory because it requires two power supplies. This defect can be remedied by building the circuit as shown in Fig. 13-2.

The Thevenin equivalent of V_{CC}, R_1, and R_2 in Fig. 13-2 is V_{BB} and R_B in Fig 13-1, where

$$V_{BB} = V_{CC} \frac{R_2}{R_1 + R_2} \quad \text{and} \quad R_B = R_1 \| R_2 \tag{13-6}$$

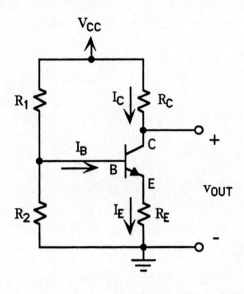

Fig. 13-2

Choosing the resistors so that $R_B \ll (\beta_F + 1)R_E$ is equivalent to making the current through R_1 and R_2 large enough so that the base current can be neglected in comparison with it. The base voltage is thus determined only by V_{CC} and the R_1, R_2 voltage divider.

The emitter resistor R_E stabilizes the operating point by providing <u>negative feedback</u>. Since the base voltage V_B is set essentially by the voltage divider, it is independent of the transistor parameters. If β_F increases for any reason, such as a temperature change, a resulting rise in I_E will increase the voltage drop across R_E, and thereby increase V_E. V_{BE} thus becomes smaller, causing a drop in I_B that counteracts the "attempted" increase in I_E.

There is no "correct" choice of component values for a bias design. The goal is to set V_{CE} so that the transistor remains in the forward-active region when the collector node swings above and below the operating point under the influence of the input signal. Therefore it is only necessary to insure that there is enough swing range so that v_C does not reach V_{CC} (cutoff), and does not fall below v_B (saturation). The required range depends on the voltage gain of the amplifier and the maximum amplitude of the input signal. In general, many combinations of component values will accomplish this goal. Bias design is typical of many engineering tasks, which have no unique solution because not enough conditions are specified. The experience, intuition, and creativity of the designer are crucial.

MOSFET Biasing

The basic-cell configuration in Fig. 13-2 is also appropriate for biasing a MOSFET inverter. The choice of resistor values is easier than for the BJT case because no gate current flows in the MOSFET. The source resistor R_E serves the same purpose as before - to render the operating point insensitive to changes in the MOSFET parameter K. The condition on the resistor values required for this insensitivity is obtained by using the Thevenin equivalents in (13-6) and writing the KVL equation for the input loop:

$$V_{BB} = V_{GS} + I_D R_E. \tag{13-7}$$

Combining this with the i-v equation for the MOSFET in the constant-current region,

$$I_D = K(V_{GS} - V_{TR})^2 \tag{13-8}$$

gives a quadratic equation in I_D.

$$I_D^2 R_E^2 - I_D \left[\frac{1}{K} + 2R_E(V_{BB} - V_{TR}) \right] + (V_{BB} - V_{TR})^2 = 0. \tag{13-9}$$

If the resistances are chosen so that

$$2R_E(V_{BB} - V_{TR}) \gg \frac{1}{K}, \tag{13-10}$$

then I_D will be insensitive to changes in K. Note that V_{BB} does not depend on the actual values of R_1 and R_2, only on their ratio.

Procedure

BJT Biasing

Your goal in this experiment is to use the circuit of Fig. 13-3 to establish a stable operating point. This circuit differs from the one in Fig. 13-2 only in the use of bipolar power supplies, which permit the designer to set the input node at or near zero volts. This is often a convenience. You are to choose values for R_1, R_2, R_C, and R_E that will set I_C at about 1 mA and V_{CE} between 3 V and 5 V. In addition, V_C should be roughly midway between V_{CC} and V_B to allow for sufficient output voltage swing when a signal is applied later. Recall that, for insensitivity to changes in β_F, the condition in equation (13-5) ($R_B \ll (\beta_F + 1)R_E$) must be satisfied.

1) Begin with the transistor that you used in Experiment 11. Energize the circuit and measure the operating-point values I_C and V_{CE}. *(Under what conditions can you measure V_{CE} by placing a voltmeter directly across the transistor terminals?)* The collector current is most easily found by measuring the voltage drop across R_C or R_E. Be sure to avoid ground loops. Insert a

microammeter between the terminals a and a' to measure I_B. Calculate β_F for this transistor and compare it with your result from Experiment 11.

2) Substitute at least two other transistors into the circuit and measure I_C, V_{CE}, and β_F for each. If all three have about the same β_F, try others until you find two with significantly different β_F values. *Verify that the operating point is indeed relatively insensitive to β_F.*

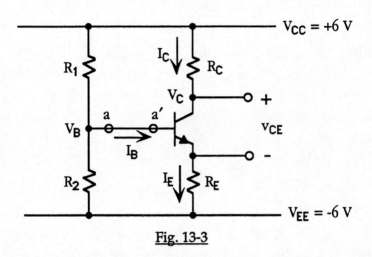

Fig. 13-3

3) Repeat part 2 after increasing R_1 and R_2 by a factor of about 50 or 100, keeping their ratio unchanged. Record your results and *compare them with your predictions as to whether the operating point remains insensitive to changes in β_F.*

4) Save your biasing circuit. This "basic cell" will become the building block for the next experiment.

MOSFET Biasing

5) Build the circuit in Fig. 13-4, using one the NMOS devices on the CD4007 chip that you measured in Experiment 9. Choose the resistor values to give $I_D \approx 1$ mA and to set V_D approximately midway between V_G and V_{DD}. Knowing the K value from your measurements in Experiment 9, try to choose the resistor values so that the stability condition in eqn (13-10) is <u>not</u> satisfied. After measuring V_D and I_D, repeat the measurement with another NMOS, which may have a different K value. *Did the operating point change significantly?*

6) Repeat the measurements of part 5 with the same NMOS devices, this time using resistor values that satisfy the stability condition. *Was there less change in the operating point this time?*

7) Save the components of this MOSFET cell for use in Experiment 15.

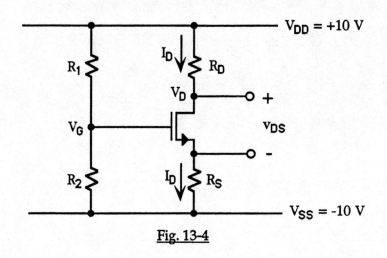

Fig. 13-4

EXPERIMENT 14

THE BJT COMMON-EMITTER AMPLIFIER

7.4 Small-Signal Modeling of Analog Circuits
    7.4.1 Incremental Signal
    7.4.2 Small-Signal Model of the BJT
    7.4.5 Transistor Small-Signal Output Resistance
    7.4.6 Transistor Small-Signal Input Resistance

Purpose

Now that the operating point of your BJT amplifier has been set, the circuit will be used in two forms; the common-emitter configuration amplifies the voltage of an input waveform, and the emitter follower reproduces the waveform without voltage amplification, but with sufficient current to drive a low-resistance load. The follower will be studied in Experiment 16. The small-signal parameters - intrinsic gain, input resistance, and output resistance - will be measured.

Introduction

We now reconsider the basic cell that was studied in Experiment 13 (Fig. 13-3). To use it as a voltage amplifier, we must first connect an input signal source and an output load to it without disturbing its operating point. In circuits composed of discrete components, as opposed to integrated-circuit chips, this goal is often accomplished with isolation capacitors, as shown in Fig. 14-1.

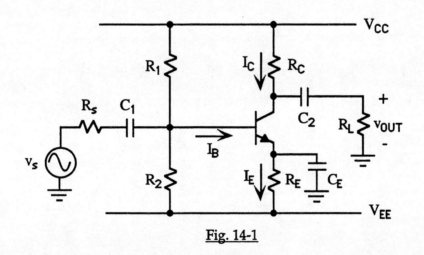

Fig. 14-1

The dc state of the circuit is determined by the resistors R_1, R_2, R_C, and R_E. The signal source, whose Thevenin equivalent is v_s, R_s, is connected through the isolation capacitor C_1, which is large enough to present a negligibly small impedance at the frequency of the signal. At the same time, C_1 acts as an open circuit at dc, so that the dc voltage at the transistor base is unaffected by the signal source. C_2, which is also large, acts in the same way to isolate the circuit from the load resistor at dc, while allowing the signal to reach the load with little attenuation. The bypass capacitor C_E serves a different function, which will be discussed later.

To provide linear amplification of an input waveform, the BJT must operate within its <u>small-signal limitation</u> (review Experiment 6), where the change in base-emitter diode current is effectively proportional to the change in base-emitter voltage, which is the input signal. We found in Experiment 6 that this limitation is $v_{be} < V_T$, the thermal voltage (0.026 V at room temperature). When this condition holds, the small-signal model of the transistor, shown in Fig. 14-2, can be used.

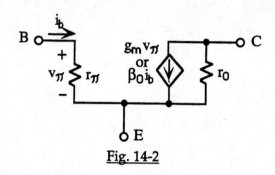

Fig. 14-2

Since this is a small-signal, or incremental model, all currents, node voltages and voltage differences denote deviations from the operating point, not total values. The incremental base-emitter voltage is called v_π here. r_π, the incremental resistance of the base-emitter diode, is the equivalent of r_d in Experiment 6. β_0 is the ratio i_c/i_b, and $r_0 = v_{ce}/i_c$. The latter accounts for the small slope of the i-v characteristics in the forward-active region. For most purposes, we will take β_0 to equal β_F.

The small-signal behavior of any BJT circuit can be derived by redrawing the circuit with all transistors replaced by Fig. 14-2, then applying the usual analysis tools to the resulting linear network. Doing this with Fig. 14-1, first with C_E removed, we get Fig. 14-3.

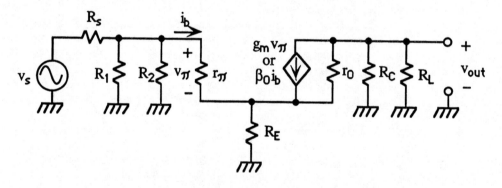

<u>Fig. 14-3</u>

Any node that has an unchanging voltage V_X in the actual circuit, whether at V_{CC}, V_{EE} or zero, can be described by the statement $\Delta V_X = 0$, or $v_x = 0$. That node thus appears at ground in the small-signal circuit. A new symbol is used for ground in the incremental circuit to emphasize that it denotes an <u>incremental, or ac ground.</u>

The linear analysis of Fig. 14-3 leads to the following expression for the voltage gain:

$$\frac{v_{out}}{v_s} \approx -\left(\frac{R_1\|R_2}{R_1\|R_2 + R_s}\right)\frac{\beta_0 R_C\|R_L\|r_0}{r_\pi + (\beta_0+1)R_E} \tag{14-1}$$

The gain is very low because of the term $(\beta_0+1)R_E$ in the denominator, which is large because β_0 is of the order of 100. The emitter bypass capacitor C_E is added in parallel with R_E to remove this term. The impedance of C_E is low enough at the signal frequency to make it appear effectively as a short circuit for R_E. After adding C_E, combining parallel resistors, and ignoring r_0, because it is in parallel with the much smaller R_C and R_L, we arrive at the version of the incremental circuit shown in Fig. 14-4.

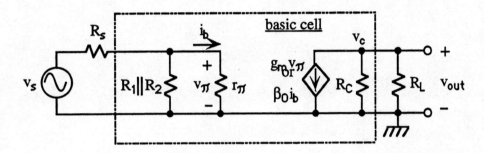

Fig. 14-4

The basic cell amplifier is inside the dashed rectangle; the Thevenin equivalent of the signal source is at the left, and the load resistor is at the right. We derive the voltage gain by first finding the Thevenin-block, or two-port equivalent of the basic cell. The intrinsic gain a_v of the cell is

$$a_v = \frac{v_c}{v_\pi} = -g_m R_C. \tag{14-2}$$

The input resistance, or Thevenin resistance, looking into the input port from the left, is

$$r_{in} = R_1\|R_2\|r_\pi. \tag{14-3}$$

The output resistance, or Thevenin resistance, looking into the output port from the right, is

$$r_{out} = R_C, \tag{14-4}$$

because, when the input source is turned off, the current from a test source at the output port flows only through R_C, since $v_\pi = 0$.

Any amplifier, no matter how complicated, can be represented by its two-port equivalent, once the three quantities a_v, r_{in}, and r_{out} are known. The Thevenin equivalent of the amplifier is shown in Fig. 14-5.

The overall gain v_{out}/v_s is

$$\frac{v_{out}}{v_s} = \left(\frac{r_{in}}{r_{in}+R_s}\right)(a_v)\left(\frac{R_L}{R_L+r_{out}}\right) = \left(\frac{R_1\|R_2\|r_\pi}{R_1\|R_2\|r_\pi+R_s}\right)(-g_m R_C)\left(\frac{R_L}{R_L+R_C}\right). \tag{14-5}$$

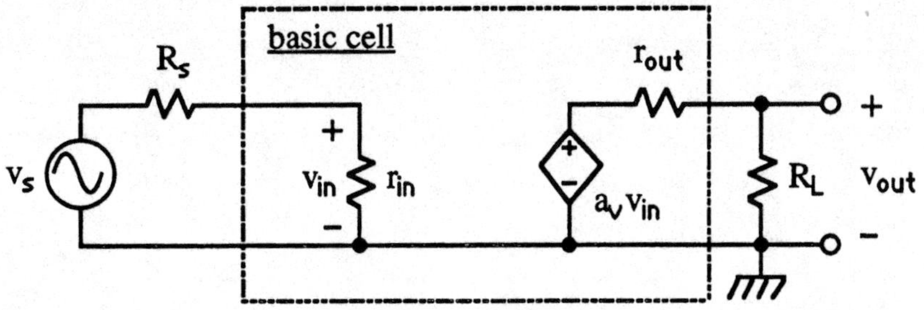

Fig. 14-5

Note that (14-5) is the same as (14-1) after the $(\beta_0+1)R_E$ term is removed (by adding C_E) and the substitution $\beta_0 = g_m r_\pi$ is made. The three terms in parentheses are, from left to right, the <u>input loading factor</u>, the <u>intrinsic gain</u>, and the <u>output loading factor</u>.

Procedure

BEFORE COMING TO THE LABORATORY

Build the amplifier circuit of Fig. 14-6, using as the nucleus your previously biased circuit from Experiment 13. The voltage divider in the signal source allows you to provide a small signal voltage to the amplifier with the amplitude control of the signal generator set near the middle of a higher range, where more precise adjustment can be achieved. Recommended values are 10 kΩ and 100 Ω for R_x and R_y, respectively. You should derive the Thevenin equivalent of this source network and use it in your analysis of the amplifier.

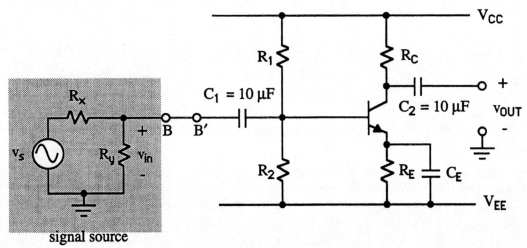

Fig. 14-6

Be sure to observe the correct polarity of any electrolytic capacitors that you use for C_E. They may explode if you don't.

IN THE LABORATORY

1) Set the frequency of the signal source to 10 kHz and set the amplitude to make v_{in} about 10 mV. You will measure v_{out} by observing the peak-to-peak values with the oscilloscope. It will be helpful to use "external triggering" of the scope, connecting the trigger input to the "trig out" terminal of the signal generator.

   a. Begin with no bypass capacitor C_E in the circuit, and record v_{out} for several different values of v_{in}.
   b. Repeat the measurements for C_E = 10 μF and 100 μF.
   c. Repeat the measurement for one value of v_{in}, with C_E = 100 μF, and with a 100-kΩ load resistor across the output terminals.
   d. Repeat part 1c with a 220-Ω load resistor across the output terminals
   e. Repeat part 1c, increasing v_{in} while observing the output waveform on the oscilloscope. Continue until the waveform becomes grossly distorted, indicating that the transistor has reached cutoff or saturation.

2) With the input set to one of the values from part 1, insert a variable resistance between terminals B and B'; adjust its value until the output voltage is halved, and record the resistance value, which is the input resistance r_{in} of the amplifier.

3) Measure the output resistance r_{out} of the amplifier by placing a variable resistor across the output terminals, again adjusting its value until the output is halved. Record the resistance value.

Keep the circuit of Fig. 14-6 intact for use in Experiment 16.

Analysis of the Data

A) Now that you have experimental values for the overall gain of the amplifier, r_{in} and r_{out}, the remaining unknown terms in (14-5) is g_m. Calculate g_m from the data and compare it with the vakue obtained from.

$$g_m = \frac{I_C}{V_T} \qquad (14\text{-}6)$$

where I_C is the operating-point current and V_T is the thermal voltage. You can also find r_π from r_{in} using (14-3), and then find β_0 for the transistor from $\beta_0 = r_\pi g_m$. Are your results consistent with values that you obtained in earlier experiments?

B) Does your observation in part 1e indicate that the transistor has gone into cutoff, or into saturation? Is the result consistent with what you would predict from your component values?

C) How do you explain the change in voltage gain in parts 1c and 1d when the load resistance was changed?

D) The resistance that you measured in part 2 is approximately r_{in}. You should draw an equivalent circuit for the amplifier with the additional resistance to convince yourself that this method for finding r_{in} is valid. Note that it requires that the Thevenin resistance of the source, $R_x \| R_y$, be small compared with r_{in}. Compare this measured value of r_{in} with the value calculated from 14-3.

E) The resistance that you measured in part 3 is approximately r_{out}. Again you should convince yourself that this method works. Compare your measured r_{out} with (14-4.)

EXPERIMENT 15

MOSFET COMMON-SOURCE AMPLIFIER

7.4 Small-Signal Modeling of Analog Circuits
    7.4.1 Incremental Signal
    7.4.3 Small-Signal Model of the FET
    7.4.5 Transistor Small-Signal Output Resistance
    7.4.6 Transistor Small-Signal Input Resistance

Purpose

To study the similarities and differences between the common-emitter amplifier and its MOSFET counterpart - the common-source amplifier.

Introduction

The common-source amplifier to be studied in this experiment is shown in Fig. 15-1.

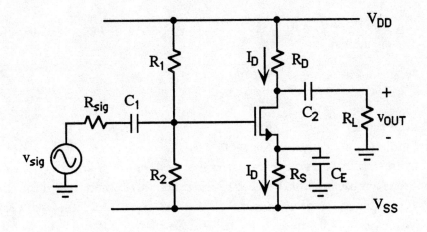

Fig. 15-1

This circuit differs from the common-emitter amplifier in Fig. 14-1 only in that an NMOS replaces the BJT, and subscripts are changed to be consistent with MOSFET terminology. The signal source is now represented by v_{sig} and R_{sig} to avoid confusion with the source resistor. The other circuit elements have the same functions as before. (You should review the introduction to Experiment 14.)

To provide linear amplification, the MOSFET must operate in its constant-current region, and also within its small-signal limitation. Since there is no forward-biased diode, the small-signal limitation is not the same as for the BJT. It can be found by writing the MOSFET equation for the total drain current as the sum of the dc operating-point value I_D and the incremental part i_d:

$$i_D = I_D + i_d = K[(V_{GS} + v_{gs}) - V_{TR}]^2$$
$$= K[(V_{GS} - V_{TR}) + v_{gs}]^2$$
$$= K[(V_{GS} - V_{TR})^2 + 2v_{gs}(V_{GS} - V_{TR}) + v_{gs}^2]$$

(15-1)

Linearity requires that

$$v_{gs}^2 \ll 2v_{gs}(V_{GS} - V_{TR}), \text{ or}$$
$$v_{gs} \ll 2(V_{GS} - V_{TR}).$$

(15-2)

Then (15-1) gives

$$I_D = K(V_{GS} - V_{TR})^2$$

(15-3)

and

$$i_d \approx 2K(V_{GS} - V_{TR})v_{gs} = g_m v_{gs}.$$

(15-4)

By substituting for (V_{GS} - V_{TR}) from (15-3), the transconductance g_m can also be written as

$$g_m = \frac{i_d}{v_{gs}} = 2K(V_{GS} - V_{TR}) = 2\sqrt{KI_D}.$$

(15-5)

These equations point out two significant differences between the BJT and the MOSFET. First, the small signal limit is much larger in the MOSFET. For V_{GS} - V_{TR} = 1 V, equation (15-2) shows that v_{gs} can be several tenths of a volt, as opposed to a limit of less than 25 mV for v_{be} in the BJT. Second, the transconductance of the MOSFET is much lower than that of the BJT for the same current. If, for example, the drain current is 1 mA in a typical MOSFET with K = 0.1 mA/V², equation (15-5) gives $g_m \approx 0.6$ mA/V. A BJT with I_C = 1 mA has g_m = 40 mA/V. Because of the resulting low gain, the simple common-source amplifier in Fig. 15-1 is not ordinarily used. Still, it is worth our attention as a means for studying the small-signal behavior of the MOSFET.

The small-signal model for the MOSFET in the forward-active region is

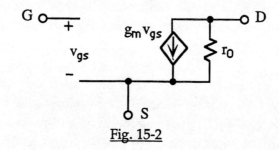

Fig. 15-2

The gate-to-source port looks like an open circuit because no current flows through the gate insulator. As in the BJT, the resistor r_0 represents the slight slope of the i-v curves in the forward-active region. The MOSFET counterpart of Fig. 14-4 for the amplifier with a bypass capacitor is

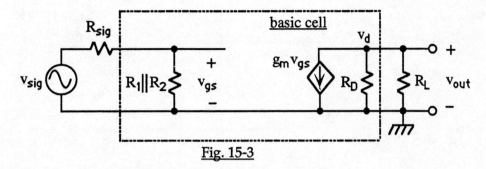

Fig. 15-3

The intrinsic gain a_v of the basic cell is

$$a_v = \frac{v_d}{v_{gs}} = -g_m R_D. \qquad (15\text{-}6)$$

The input resistance is

$$r_{in} = R_1 \| R_2, \qquad (15\text{-}7)$$

and the output resistance is simply R_D. Comparing with (14--5), the overall gain v_{out}/v_{in} for the MOSFET amplifier is

$$\frac{v_{out}}{v_s} = \left(\frac{r_{in}}{r_{in} + R_{sig}}\right)(a_v)\left(\frac{R_L}{R_L + r_{out}}\right) = \left(\frac{R_1 \| R_2}{R_1 \| R_2 + R_{sig}}\right)(-g_m R_D)\left(\frac{R_L}{R_L + R_D}\right). \qquad (15\text{-}8)$$

Without the source bypass capacitor, the voltage gain will be much lower, just as it was in the BJT case.

Procedure

BEFORE COMING TO THE LABORATORY

Build the amplifier circuit of Fig. 15-4, using as the nucleus your previously biased MOSFET circuit from Experiment 13. The voltage divider in the signal source allows you to provide a small signal voltage to the amplifier with the amplitude control of the signal generator set near the middle of a higher range, where more precise adjustment can be achieved. Recommended values are 10 kΩ and 100 Ω for R_x and R_y, respectively. You should derive the Thevenin equivalent of this source network and use it in your analysis of the amplifier.

Be sure to observe the correct polarity of any electrolytic capacitors that you use for C_E. They may explode if you don't.

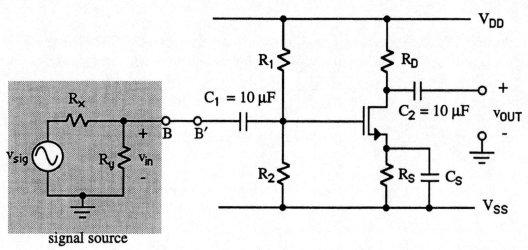

Fig. 15-4

IN THE LABORATORY

1) Set the frequency of the signal source to 10 kHz and set the amplitude to make v_{in} about 10 mV. You may find that a larger value is needed when you begin your measurements. Remember that the voltage gain of this amplifier will be lower than what you observed in the common-emitter amplifier. because of thesmaller g_m. You will measure v_{out} by observing the peak-to-peak values with the oscilloscope. It will be helpful to use "external triggering" of the scope, connecting the trigger input to the "trig out" terminal of the signal generator.

    a. Begin with no bypass capacitor C_E in the circuit, and record v_{out} for several different values of v_{in}.
    b. Repeat the measurements for C_E = 10 µF and 100 µF.
    c. Repeat the measurement for one value of v_{in}, with C_E = 100 µF, and with a 100-kΩ load resistor across the output terminals.
    d. Repeat part 1c with a 220-Ω load resistor across the output terminals.
        Continue until the waveform becomes grossly distorted, indicating that the transistor has reached cutoff or the triode region.

2) With the input set to one of the values from part 1, insert a variable resistance between terminals B and B'; adjust its value until the output voltage is halved, and record the resistance value, which is the input resistance r_{in}.

3) Measure the output resistance r_{out} of the amplifier by placing a variable resistor across the output terminals, again adjusting its value until the output is halved. Record the resistance value.

Analysis of the Data

A) Now that you have an experimental value for the overall gain of the amplifier, the remaining unknown term in (15-8) is g_m. Compare the values of g_m calculated from (15-8) and (15-5). You will need the K value that you measured in Experiment 9.

B) Does your observation in part 1e indicate that the transistor has gone into cutoff, or into the triode region? Is the result consistent with what you would predict from your component values?

C) How do you explain the change in voltage gain in parts 1c and 1d when the load resistance was changed?

D) The resistance that you measured in part 2 is approximately r_{in}. If you have not already done so in Experiment 14, you should draw an equivalent circuit for the amplifier with the additional resistance to convince yourself that this method for finding r_{in} is valid. Note that it requires that the Thevenin resistance of the source, $R_x \| R_y$, be small compared with r_{in}. Compare this measured value of r_{in} with the value calculated from (15-7).

E) The resistance that you measured in part 3 is approximately r_{out}. Again you should convince yourself that this method works, if you have not done so in Experiment 14. Compare your measured r_{out} with your predicted value.

EXPERIMENT 16

THE BJT EMITTER FOLLOWER

6.1.1 BJT Inverter
6.2.1 BJT Voltage Follower

Purpose

The emitter follower will be studied and used as a buffer in combination with the common-emitter amplifier to drive a low-resistance load with minimal loss in voltage amplification. You will find that when the common-emitter amplifier is used alone with a low-resistance load, the overall voltage gain is much lower than the intrinsic gain because of a reduction in the output loading factor. Adding an emitter follower restores much of the missing gain.

Introduction

Figure 16-1 shows a simple form of the emitter follower. The output is measured at the emitter of the transistor, and no collector resistor is needed. The resistors R_1, R_2, and R_E are used as before to set the operating point, and the coupling capacitors provide dc isolation from the signal source and the load. R_s is the Thevenin resistance of the signal source.

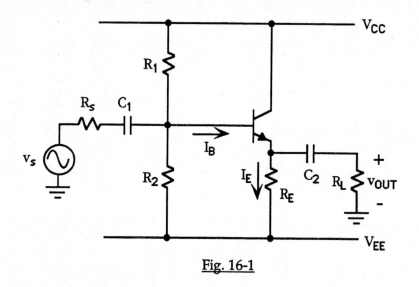

Fig. 16-1

The Thevenin-block quantities for the amplifier are:

$$r_{in} = R_1 \| R_2 \| (r_\pi + (\beta_0 + 1)R_E), \qquad (16\text{-}1)$$

$$r_{out} = R_E \left\| \frac{r_\pi + R_1 \| R_2 \| R_s}{\beta_0 + 1} \right., \qquad (16\text{-}2)$$

and
$$a_v = \frac{(\beta_0+1)R_E}{r_\pi + (\beta_0+1)R_E}. \qquad (16\text{-}3)$$

r_{out} was obtained by turning off v_s, removing R_L, connecting a test source across the output terminals, and finding the expression for v_{test}/i_{test}. The intrinsic gain is obviously less than unity, although not far from it because $(\beta_0+1)R_E \gg r_\pi$.

The usefulness of the emitter follower arises from its small output resistance, which is usually less than 100 Ω, and its fairly large input resistance, of the order of 100 kΩ or more. When the follower is inserted between the output port of the common-emitter amplifier and the load, it acts as a <u>buffer</u>; even though it contributes no further voltage gain, it prevents "loading down" of the amplifier by the load resistance. Stated another way, it provides <u>current gain</u>, providing enough current to develop the necessary voltage drop across a low-resistance load. The voltage gain of the combined stages can be found from the Thevenin circuit in Fig. 16-2.

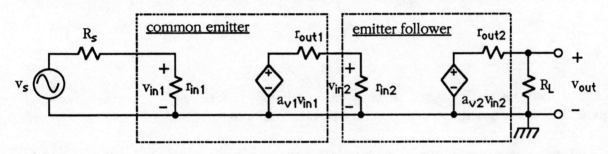

Fig. 16-2

A new term, the <u>interstage loading factor</u>, given by $\dfrac{r_{in2}}{r_{in2}+r_{out1}}$, is introduced by insertion of the follower. It is not much less than unity because the term $(\beta_0+1)R_E$ in (16-1) makes r_{in2} fairly large. Similarly, the output loading factor $\dfrac{R_L}{R_L + r_{out2}}$ is not very small because the (β_0+1) term in the denominator of (16-2) makes r_{out2} small.

An important generalization follows from this discussion: **a voltage amplifier should have as large an input resistance and as small an output resistance as possible.**

Procedure

BEFORE COMING TO THE LABORATORY

Build the circuit in Fig. 16-1, using the following component values:
- $R_1 = 22\ k\Omega$
- $R_2 = 22\ k\Omega$
- $R_E = 6.8\ k\Omega$
- $R_s = 1\ k\Omega$
- $C_1 = C_2 = 10\ \mu F$
- $V_{CC} = -V_{EE} = 6\ V$.

Calculate the operating-point current I_C, and r_{in}, r_{out}, and a_v, using equations 16-1, 16-2, and 16-3. Recall that

$$r_\pi = \frac{\beta_0}{g_m} = \frac{\beta_0 V_T}{I_C}. \qquad (16\text{-}4)$$

IN THE LABORATORY

1) Using the oscilloscope to measure the output voltage of circuit 16-1, as you did in Experiment 14, measure the voltage gain for a 10-kHz, 10-mV input signal, and load resistors of 100 kΩ and 220 Ω.

2) Measure r_{in} and r_{out} of the follower, using the same method you used in Experiment 14.

3) Combine the circuits of Fig. 14-6 and Fig. 16-1, as shown below in Fig. 16-3.

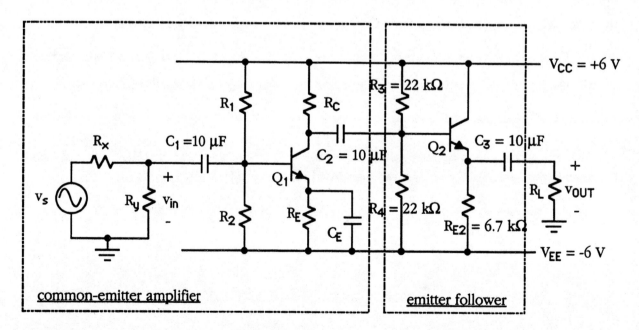

Fig. 16-3

Note that each stage is isolated by capacitors, so that its operating point can be set independently. Repeat the measurements of part 1, tabulating the measured voltage gains for both values of load resistor.

Analysis of the Data

A) Compare your measured values of a_v, r_{in}, and r_{out} for the follower with values calculated from equations (16-1), (16-2), and (16-3).

B) Calculate the overall gain of the two-stage amplifier in part 3, using the Thevenin-block equivalent model (Fig. 16-2). Compare the calculated and measured voltage gains with the 100-kΩ and 220-Ω load resistors for the common-emitter amplifier alone (Experiment 14) and for the two-stage amplifier in part 3. Did the addition of the follower stage accomplish the expected improvement in voltage gain?

Design Project

Suppose you wish to use your two-stage amplifier in Fig. 16-3 with a signal source whose Thevenin resistance is 10 kΩ instead of the 100 Ω you used in this experiment. A calculation similar to the one you did in the preceding analysis with your low-output-resistance source shows that the overall voltage gain will be considerably less than what you observed in this experiment. The reason for the decrease is a reduction in the input loading factor. Design on paper an extension of the two-stage circuit that will increase the input loading factor, and consequently the overall gain, when the 10-kΩ source is used. No increase in intrinsic gain is needed.

EXPERIMENT 17

BJT CURRENT MIRRORS

8.3.7 BJT Current Mirror
8.3.8 BJT Widlar Current Source

Purpose

To study various forms of the current mirror, which is used as a current source for biasing circuits on integrated-circuit chips.

Introduction

The current mirror is commonly used to provide bias current on analog integrated-circuit chips because it uses few components, and therefore requires little chip area. In the current mirror shown in Fig. 17-1, all the transistors are <u>matched</u>, in that they have the same value of β and the same size, so that their collector currents are equal for equal v_{BE}. The reference current I_{ref} is determined by the resistors R_{ref} and R_A in series with the diode-connected transistor Q_A.

$$I_{ref} \approx \frac{0 - (V_{EE} + 0.7)}{R_{ref} + R_A}, \text{ ignoring the base currents.} \tag{17-1}$$

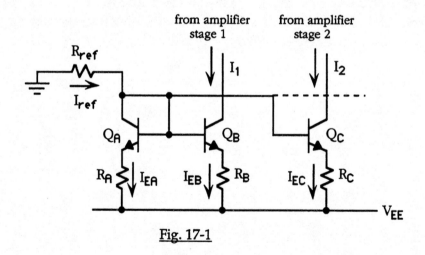

Fig. 17-1

The relation between I_1 and I_{ref} is obtained by starting with KVL around the lower loop.

$$-I_{EA}R_A - V_{BEA} + V_{BEB} + I_1 R_B = 0 \tag{17-2}$$

Using the exponential diode equation with $\eta \approx 1$ to relate I_E to V_{BE}, and using $I_{EA} \approx I_{ref}$, gives

$$I_1 \approx \frac{V_T}{R_B} \ln \frac{I_{ref}}{I_1} + I_{ref} \frac{R_A}{R_B}, \tag{17-3}$$

where V_T is the thermal voltage (0.026 V at room temperature). Equation (17-3) is the defining relationship for this form of the current mirror, known as a <u>Wilson source</u>. If I_{ref} and I_1 are not very different, the logarithmic term is close to zero, and can be neglected, giving

$$\frac{I_1}{I_{ref}} \approx \frac{R_A}{R_B}. \tag{17-4}$$

In this case it is not necessary that the two transistors be closely matched in their β values or their scale currents. In principle, both R_A and R_B could be removed, giving $I_1 \approx I_{ref}$ for matched transistors. However, the use of the resistors is preferable because they provide negative feedback that stabilizes the current. As shown in the diagram, the same reference structure can provide bias current I_2 to another amplifier stage with the addition of only one transistor and one resistor. In practice, one reference structure is used for two or three stages on a chip.

Another variation on the current mirror is the <u>Widlar source</u>, which is used when a very low bias current is needed. It is obtained by removing R_A from Fig. 17-1. The relationship between I_1 and I_{ref} then becomes

$$I_1 = \frac{V_T}{R_B} \ln \frac{I_{ref}}{I_1}. \tag{17-5}$$

If, for example, a bias current $I_1 = 20$ μA is needed with $V_{EE} = -15$ V, the Wilson source would require a resistance $R_{ref} + R_A$ of about 1 megohm, which occupies a prohibitively large amount of chip area. On the other hand, a Widlar source with $I_{ref} = 1$ mA provides 20 μA with $R_{ref} = 14.3$ kΩ and $R_B = 5$ kΩ. The reference current of 1 mA can be used to bias other circuits on the chip. The reason such a large reduction in current is achieved with a relatively small resistor is that a small voltage drop in the emitter branch of Q_B reduces V_{BE} of Q_B by a small amount., which in turn causes a large decrease in I_1 because of the exponential i_B-v_{BE} relationship of the transistor.

For a current mirror to approximate an ideal current source, its incremental resistance r_{cs}, looking into the collector of the output transistor, should be as large as possible. The simple current mirror with no emitter resistors has an output resistance of r_0, the collector resistance of the transistor (around 100 kΩ). A desirable feature of the Widlar source is that addition of the small R_B increases the incremental output resistance of the source considerably. The Widlar source has a resistance given by

$$r_{cs} \approx r_0 \left(1 + \frac{\beta_0 \ln \frac{I_{ref}}{I_1}}{\beta_0 + \ln \frac{I_{ref}}{I_1}} \right). \tag{17-6}$$

For the example discussed above, where $I_{ref} = 50 I_1$, $r_{cs} = 4.8 r_0$. The Wilson source behaves similarly.

Procedure

1) Build the Wilson source in Fig.17-2, with $R_A \approx 2$ kΩ and a 10-kΩ potentiometer connected as a variable resistor for R_B. Use a curve tracer or some other appropriate measurement tool to select transistors with approximately equal values of β_F. Choose the value of R_{ref} to give $I_{ref} \approx 1$ mA.

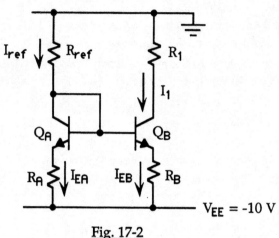

Fig. 17-2

2) Measure I_{ref} either with an ammeter or with the oscilloscope connected to measure the voltage across R_{ref} (be careful about grounding). Then measure I_1 in the same way for several values of R_B, selecting values that make I_1 both less than and greater than I_{ref}. Measure R_B directly for each value of I_1. R_1 should be large enough to give an easily measurable voltage drop across it with the scope, but should not be so large that Q_B is forced into saturation. Note that the oscilloscope can be used for the current measurement because both R_{ref} and R_1 have one terminal connected to ground.

3) Repeat the measurements of part (a) with R_A removed. You now have a Widlar source.

4) If the r_0 of your transistors is not too large, you may be able to measure the incremental resistance r_{cs} of the current source when looking into the collector of Q_B. Try to make this measurement by connecting the top of R_1 to the positive terminal of a voltage source instead of to ground. Them measure I_1 with a milliammeter for different values of the positive voltage.

5) Label your matched pair of transistors so you can use them when you build a differential amplifier.

Analysis of the Data

A) Compare the I_1 vs R_A data of parts 2 and 3 with the values calculated from (17-3) and (17-5), respectively. Over what range of I_1 can you use the approximate relationship in (17-4) without appreciable error?

B) If you were able to obtain a meaningful value of r_{cs} in part 4, compare it with the value calculated from (17-6).

EXPERIMENT 18

THE BJT DIFFERENTIAL AMPLIFIER

8.1 Basic Differential Amplifier Topology
8.2 Differential and Common-Mode Signals
8.3 BJT Differential Amplifier
    8.3.1 BJT Differential Amplifier with One Input
    8.3.2 BJT Differential Amplifier with Two Inputs
    8.3.4 Common-Mode Rejection Ratio
    8.3.6 BJT Differential Amplifier Biasing
    8.3.7 BJT Current Mirror

Purpose

The differential amplifier, which generates an output signal proportional to the difference between two independent input signals, is an essential component of many analog integrated circuits. In particular, it is the first stage of all operational amplifiers. You will study its characteristics in this experiment.

Introduction

In all the previous experiments in this manual, the circuits were constructed using the rules for discrete-circuit design. Any component that was available in discrete form could be used. Most modern circuits are, however, built as an <u>integrated circuit</u> (IC) on a single chip of silicon. Integrated circuits have lower cost, higher speed, lower power dissipation, and greater reliability than discrete circuits. However, one disadvantage of the integrated circuit that must be addressed is the limitation in the types and sizes of components that can be used. This leads to the following practices:

Capacitors are limited to a maximum of about 50 pF because larger ones would occupy too large an area on the silicon chip. This precludes the use of capacitors for dc isolation between amplifier stages or for bypassing emitter resistors.

Large-value resistors are avoided because of the space they occupy, and the number of resistors is kept to a minimum for the same reason. On the other hand, the designer is not limited to the use of the manufacturer's stock values for resistors, and resistor tolerances can be made small. This allows the liberal use of "matched pairs" of resistors.

Since transistors occupy little chip area, they are used more liberally than they are in discrete circuits. Furthermore, because transistors made on the same chip and separated by only a few micrometers from each other have almost identical values of β and cross-sectional area, matched pairs of transistor are available, and are used frequently. It is time-consuming and expensive to match discrete transistors.

Using the guidelines discussed above to design an integrated-circuit version of the common-emitter amplifier leads to the circuit of Fig. 18-1.

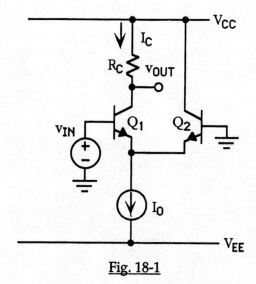

Fig. 18-1

The operating-point current I_0 is not set by a resistive voltage divider, but by the current source, which, in integrated circuits, is usually a current mirror such as the ones studied in Experiment 17.

Assume for the moment that Q_2 is not present. The incremental resistance r_{cs} of the current source acts as an emitter resistance for Q_1, yielding the voltage gain expression

$$\frac{v_{out}}{v_{in}} = \frac{-\beta_0 R_C}{r_\pi + (\beta_0 + 1)r_{cs}} \qquad (18\text{-}2)$$

This is much less than unity because r_{cs} is so large. Thus, an emitter bypass is essential; but it cannot be a 100-µF capacitor because "integrated-circuit rules" do not allow the use of such a large capacitor. The IC designer makes use of the fact that the incremental resistance of a BJT, "looking into" the emitter, is

$$r_e = \frac{r_\pi}{\beta_0 + 1} = \frac{\beta_0}{g_m(\beta_0 + 1)} \approx \frac{1}{g_m} \qquad (18\text{-}3)$$

For an operating-point current of the order of 1 mA, $r_e \approx 25$ ohms - a relatively small resistance. This is the reason for using the matching transistor Q_2 as the bypass element. Since the emitters of Q_1 and Q_2 are connected together and their bases are grounded at the operating point, V_{BE} is the same for both; therefore their collector currents are the same, and equal to $I_0/2$, and they have the same value of r_π. Note that the collector current is not determined by the size of the collector resistor, but by V_{BE} of the transistor.

The incremental representation of the amplifier in midband is shown in Fig. 18-2. From this circuit we get the equations

$$v_{out} = -g_m v_\pi R_C = -\beta_0 i_b R_C, \text{ (ignoring the larger } r_{01} \text{ relative to } R_C.)$$

$$i_b = \frac{v_{in}}{r_\pi + (\beta_0+1)r_{cs}\|r_e} \approx \frac{v_{in}}{r_\pi + (\beta_0+1)\frac{r_\pi}{\beta_0+1}} = \frac{v_{in}}{2r_\pi}$$

$$\therefore \frac{v_{out}}{v_{in}} \approx \frac{-\beta_0 R_C}{2r_\pi} = \frac{-g_m R_C}{2}$$

(18-4)

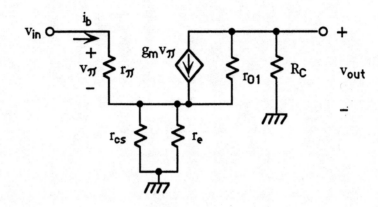

Fig. 18-2

Equation (18-4) shows that the voltage gain with the <u>active bypass</u> Q_2 is one-half the gain of an amplifier that uses a large bypass capacitor. This is the price that must be paid for replacing the large capacitor with the active bypass.

The differential amplifier of Fig. 18-3 is created when the circuit of Fig. 18-1 is made symmetrical. A second collector resistor equal to R_C is added, and the base of Q_2 is used as a second input terminal. Now each transistor can serve as either the input element or the active bypass. The output voltage can be measured from the collector of Q_1 to ground, from the collector of Q_2 to ground, or between the two collectors.

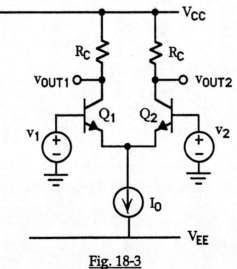

Fig. 18-3

Although the two input voltages v_1 and v_2 are independent of each other, they can be expressed as linear combinations of a <u>differential input</u> v_{idm} and a <u>common-mode input</u> v_{icm}, defined by

$$v_{idm} = v_1 - v_2 \quad \text{and} \quad v_{icm} = \frac{v_1 + v_2}{2} \qquad (18\text{-}5)$$

Rearranging these equations gives

$$v_1 = v_{icm} + \frac{v_{idm}}{2} \quad \text{and} \quad v_2 = v_{icm} - \frac{v_{idm}}{2} \qquad (18\text{-}6)$$

The output voltage for any arbitrary combination of v_1 and v_2 can be found once the voltage-gain expressions for a pure differential input ($v_{icm} = 0$) and a pure common-mode input ($v_{idm} = 0$) are known. From (18-5) it follows that pure differential input means $v_1 = -v_2$, and pure common-mode input means $v_1 = v_2$. The gain expressions are derived from the usual linear analysis of the small-signal circuit. The gains for the different cases are:

single-ended output at terminal v_{o1} (measured between the terminal and ground):

$$A_{dm\text{-}se1} = -\frac{g_m R_C}{2} \quad \text{and} \quad A_{cm\text{-}se1} = \frac{-\beta_0 R_C}{r_\pi + 2(\beta_0 + 1)r_{cs}}. \qquad (18\text{-}7)$$

single-ended output at terminal v_{o2} (measured between the terminal and ground):

$$A_{dm\text{-}se2} = \frac{+g_m R_C}{2} \quad \text{and} \quad A_{cm\text{-}se2} = \frac{-\beta_0 R_C}{r_\pi + 2(\beta_0 + 1)r_{cs}}, \qquad (18\text{-}8)$$

differential output (measured between terminals v_{o1} and v_{o2}):

$$A_{dm\text{-}diff} = A_{dm\text{-}se1} - A_{dm\text{-}se2} = -g_m R_C \qquad (18\text{-}9)$$
$$A_{cm\text{-}diff} = A_{cm\text{-}se1} - A_{cm\text{-}se2} = 0 \qquad (18\text{-}10)$$

Since the function of the differential amplifier is to enhance the difference between the inputs and to suppress the common-mode component, the figure of merit called the <u>common-mode rejection ratio</u> (CMRR), defined by

$$CMRR = 20 \log \frac{|A_{dm}|}{|A_{cm}|}, \qquad (18\text{-}11)$$

should be as large as possible. Equations (18-9) and (18-10) are true only if the two collector resistors are exactly equal.

For pure differential-mode input, the emitters of Q_1 and Q_2 are at incremental ground potential because the two inputs are equal and opposite. One can then view the circuit as two identical common-emitter amplifiers with grounded emitters. The high-frequency poles for differential input are found in the same way as they are for the common-emitter amplifier in Experiment 19.

Procedure

BEFORE COMING TO THE LABORATORY

Choose values for all the resistors in the circuit of Fig. 18-4 so that the bias current I_0 through Q_B is approximately 2 mA, and V_{CE} for Q_1 and Q_2 is approximately 6 V when $V_{CC} = -V_{EE} = 15$ V. R_A and R_B should be approximately equal and around 2 kΩ. Calculate the expected voltage gains for the different input modes and output modes, using a reasonable value for the incremental resistance r_{cs}.

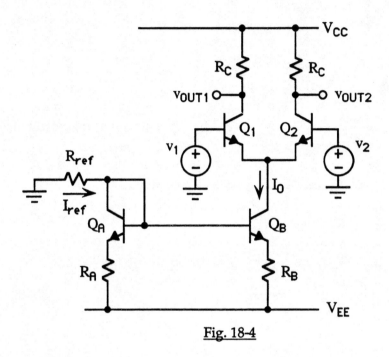

Fig. 18-4

IN THE LABORATORY

1) Using a Tektronix curve tracer or similar instrument, find two pairs of 2N2222 npn transistors, or a similar type, that have matched characteristics. You may have to exchange transistors with other students to find good matches. When measuring the transistors, take note of the value of r_o, the open-collector output resistance, given by the slope of the i_C vs v_{CE} curve in the forward-active region in the neighborhood of the expected bias point.

2) Build the circuit of Fig. 18-4 and energize it. Measure the bias currents and voltages with both input nodes grounded, and compare your measured and predicted values.

3) Apply a "single-ended" input voltage to input 1, leaving input 2 grounded. In this case, the circuit may be considered as a common-emitter amplifier with active bypass Q_2 (equation (18-4)). It may be necessary to use an input attenuator network as you did in several earlier experiments. Measure the single-ended output voltages at both of the output terminals, and compare your measured and predicted values.

4) Measure the low and high-frequency rolloff points, which are the frequencies where the gain is reduced by 3 dB, or a factor of $1/\sqrt{2}$ from its maximum value. This will require a variable-frequency signal generator with constant output amplitude. You will find that the output is independent of frequency over some frequency range, called the midband. Extend the range until the output falls to at least half the midband value.

It is difficult to apply a pure differential input to your amplifier because most signal generators provide only a single-ended output, and not a pair of signals of opposite polarity. Nevertheless, you can still evaluate the differential-mode gain by using the superposition equations (18-5) and (18-6). Given that $v_1 = v_{in}$, and $v_2 = 0$, find expressions for the differential and common modes of these signals. Express v_1 and v_2 as linear combinations of v_{idm} and v_{icm}. Measure the differential-mode output by attaching a scope probe to the collector of each transistor, and then switching the scope to the differential mode [ADD - CHAN2 INVERT]. Be sure that the two scope channels are set to the same volts/div value. In this mode, the scope will display a single trace equal to the difference between the signals on channel 1 and channel 2. *What is your measured value of $A_{dm\text{-}diff}$?*

5) Apply a pure common-mode input to your amplifier by connecting the signal generator simultaneously to both input terminals. Measure the common-mode output of the amplifier, and also measure, with a single scope probe, the voltage between the common emitter node and ground in response to the purely common-mode signal. *Find the value of A_{cm}. What is the value of the common-mode rejection ratio (CMRR) for this amplifier?*

6) Again drive the amplifier in single-ended mode (one input grounded), and increase the amplitude of v_{in} until the output clips, or limits. Determine which transistor, Q_1 or Q_2, is the first to operate outside the active region, and determine whether it goes into cutoff or into saturation. when the output voltage just begins to clip.

Analysis of the Data

A) Explain any major discrepancies between your predicted and measured values. Remember that you should use the measured values of all resistors in your circuit, rather than the values indicated by the color-code bands. These may differ by as much as ±20 percent.

B) Calculate the incremental resistance of the current mirror, looking into the collector of Q_B, from the measured common-mode gain. Compare this with the value calculated from equation (17-6).

C) Predict the behavior of your amplifier if you make one of the collector resistors twice the size of the other. Consider the dc situation, the voltage gains, and what happens to the transistors Q_1 and Q_2 when the circuit is driven as it was in part 6 of the experimental procedure. You should consider what happens when the input voltage swings both positively and negatively.

Design Project

In this experiment you built and analyzed a circuit that generates an output signal proportional to the <u>difference</u> of two inputs. A circuit that uses a current mirror for biasing can be designed to give an output proportional to the <u>product</u> of two inputs. Beginning with a circuit such as the one in Fig. 18-4, design, build, and analyze a circuit that generates the product of two different dc input voltages. The key to the design is to make use of the current mirror as part of a signal path.

Predict the limitations on input signal swing for your circuit, and predict the form of the output if the inputs are sinusoidal voltages with the same frequency and phase, and also with different frequencies. You might wish to do experimental tests of some of your predictions.

EXPERIMENT 19

FREQUENCY RESPONSE OF THE COMMON-EMITTER AMPLIFIER

9.2 Sinusoidal Steady-State Amplifier Response
    9.2.2 Bode Plot Representation in the Frequency Domain
    9.2.3 High, Low, and Midband Frequency Limits
9.3 Frequency Response of Circuits Containing Capacitors
    9.3.1 High- and Low-Frequency Capacitors
    9.3.2 The Dominant-Pole Concept
    9.3.4 Miller's Theorem and Miller Multiplication
    9.3.5 Frequency Response with Bypass Capacitor

Purpose

To determine the limits to the range of signal frequencies over which the voltage gain of the common-emitter amplifier is effectively constant.

Introduction

The voltage-gain measurements on the common-emitter amplifier and emitter follower made in Experiments 14 and 16 were done with a 10-kHz signal. At this frequency the 10-µF and 100-µF capacitors in the circuits have impedances in the range of 1 to 10 ohms - low enough for them to be treated as effective short circuits. Furthermore, the capacitances internal to the transistors are sufficiently low for them to be treated as open circuits. The 10-kHz frequency consequently falls in <u>midband</u>, where the voltage gain is essentially independent of frequency.

The gain decreases when the signal frequency is reduced to a point where the capacitors C_1, C_2, and C_E introduce an appreciable impedance. These <u>low-frequency capacitances</u> add either a series impedance in the signal path or an impedance in parallel with R_E, which is then no longer fully bypassed. A rough estimate of the low-frequency breakpoint ω_L, below which the gain is reduced, can be made with the following procedure:

    a. Consider each low-frequency capacitor separately, short-circuiting the others.
    b. Find the Thevenin resistance R_{Th} seen by the remaining capacitor (after killing the signal source).
    c. The pole frequency that is introduced by that capacitor is $1/CR_{Th}$.
    d. Repeat steps a through c for each of the remaining low-frequency capacitors.
    e. If one of the pole frequencies is at least four times the next lowest one, it is dominant, and is approximately equal to ω_L. If there is no dominant pole, then ω_L is approximately the sum of all the pole frequencies.

The low-frequency small-signal circuit is shown in Fig. 19-1.

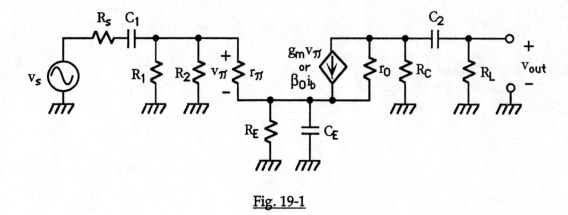

Fig. 19-1

Following the above procedure, the three low-frequency poles are

$$\omega_{p1} = \frac{1}{C_1\left[R_s + (R_1\|R_2\|r_\pi)\right]}$$

$$\omega_{p2} = \frac{1}{C_2(R_L + R_C\|r_0)}$$

$$\omega_{pE} = \frac{1}{C_E \dfrac{R_E\|(r_\pi + R_1\|R_2\|R_s)}{\beta_0 + 1}} \qquad (19\text{-}1)$$

The pole associated with C_E is usually dominant because of the large β_0+1 divisor in the denominator.

The high-frequency breakpoint ω_H is determined by the junction capacitances inside the transistor. As shown in Fig. 19-2, C_π, the base-emitter junction capacitance, shunts r_π, thereby reducing v_π. C_μ, the base-collector capacitance, feeds some of the input current to the output, subtracting from the current through r_π. Thus, both capacitances cause the gain to be reduced as the signal frequency is increased. C_μ can be transformed with the use of the Miller theorem to a capacitance

$$C_A = C_\mu(1 + g_m R_C\|R_L) \qquad (19\text{-}2)$$

in parallel with C_π. The Thevenin resistance seen by the combined capacitors is found after killing the signal source, and the high-frequency pole becomes

$$\omega_p = \frac{1}{(C_\pi + C_A)\left[r_\pi\|(r_x + R_1\|R_2\|R_s)\right]} \qquad (19\text{-}3)$$

This equals ω_H because it is the only high-frequency pole. r_x, the parasitic base resistance, which is of the order of 10 Ω, is negligible at low frequency, but can be significant at high frequency if the source resistance R_s is small. Since your amplifier (Fig. 14-6) has a source resistance of about 100 Ω, the effect of r_x is small.

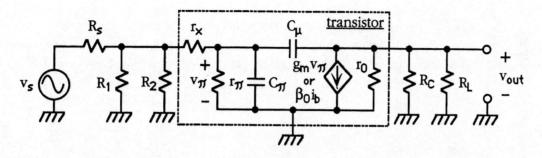

Fig. 19-2

Procedure

BEFORE COMING TO THE LABORATORY

Calculate the expected values of ω_L and ω_H for the common-emitter amplifier that you used in experiment 14 (Fig. 14-6). You will need information from the manufacturer's data sheet for the transistor to obtain a value for C_π. Typical values for the 2N2222 transistor, for example, are

$C_\mu = 8$ pF (given as C_{ob} in the data sheet)

f_T (unity-gain bandwidth) = 250 MHz.

C_π is obtained from the equation

$$C_\pi \approx \frac{g_m}{2\pi f_T} - C_\mu = \frac{\beta_0}{2\pi f_T r_\pi} - C_\mu \tag{19-4}$$

Use the value for β_0 that you determined earlier.

IN THE LABORATORY

1) Check the operating point of your amplifier to be sure that it is performing properly. The bypass capacitor C_E should be 100 μF. Use a 5-kHz input signal of constant amplitude within the small-signal limit, and measure the output voltage amplitude on the oscilloscope. Confirm that 5 kHz lies within midband by measuring the output voltage at other frequencies in the neighborhood of 5 kHz. Display both the input and the output signals on the scope with different vertical sensitivity settings so that you can measure the phase difference between the two.

2) Reduce the frequency, keeping the input amplitude constant, and measure the output voltage and the phase shift at each frequency, until the output reaches about 10 percent of the midband value. If you go low enough in frequency, you may observe the effect of the zero introduced by C_E, whose value is

$$\omega_z = \frac{1}{C_E R_E} \tag{19-5}$$

3) If a high-frequency signal generator is available, repeat the gain measurements, increasing the frequency until the gain is reduced to about 10 percent of the midband value. Again measure the phase shifts. If a generator is not available, you can use the step-response method to find ω_H. Under certain assumptions, the output waveform resulting from a step-function input is given by

$$\frac{v_{out}}{v_{in}} = A(1 - \exp(-\omega_H t)), \tag{19-7}$$

where A is the midband gain. The effects of C_E and C_1 are negligible in this case. Find the time constant $1/\omega_H$ for the exponential decay by applying a 10-kHz square-wave input to the amplifier, and observing the time required for the output voltage to reach 63 percent of its final value *(why 63 percent?)*.

4) Display the overall performance of the amplifier by applying a 2-kHz square-wave input with a sufficiently low amplitude so that it is linearly amplified. Make careful sketches of the input and output waveforms in your notebook.

Analysis of the Data

A) Using the data from parts 1 and 2 (and part 3 if available), plot the voltage gain in dB, with zero dB representing unity gain, versus ω on a logarithmic scale. Remember that $\omega = 2\pi f$, and that

$$\text{gain in dB} = 20 \log_{10} \frac{v_{out}}{v_{in}} \tag{19-6}$$

Also plot the phase differences expressed in degrees versus ω on semi-logarithmic graph paper.

B) Compare your measured values of ω_L and ω_H with your values calculated from (19-1) and (19-3), and try to explain any discrepancies.

C) Compare the shapes of the input and output waveforms obtained in part 4. How do you explain what you see? If your output doesn't look like a perfect square wave (it shouldn't), describe how you might reduce the distortion, but keep a constant output amplitude, given by the product of the input amplitude and the midband gain.

EXPERIMENT 20

TRANSIENT RESPONSE OF THE PN JUNCTION DIODE

Purpose

To measure the response times of the pn-junction diode. The <u>turn-on</u> time is the time required for steady state to be reached when the diode is switched into forward bias by a voltage step. The <u>turn-off</u> time is the time required for steady state to be reached when the diode is switched into reverse bias.

Introduction

When a pn-junction diode is forward biased, minority electrons are diffusing into the p region, and minority holes are diffusing into the n region. Current is observed when these "excess" carriers leave the diode and enter the external circuit. When the diode is suddenly switched into forward bias, time is required for the excess carriers to reach their steady-state values for two reasons: a finite time is needed for the first carriers to travel from the junction to the opposite end of the diode, and some of them are lost by recombination with the majority carriers along their path.

The situation is similar when the diode is suddenly switched from forward bias into reverse bias. The current does not go to zero instantaneously because time is needed for the excess carriers to disappear, either by recombination, or by being swept back to the opposite side of the junction.

When the diode is forward biased, the amount of excess charge Q_p due to minority holes on the n side, and the current i_p contributed by these holes, are given by

$$Q_p = K_1\left(e^{\frac{v_D}{\eta V_T}} - 1\right) \text{ and } i_p = K_2\left(e^{\frac{v_D}{\eta V_T}} - 1\right), \tag{20-1}$$

where the constants K_1 and K_2 are functions of the equilibrium hole concentrations and the dynamics of excess hole motion on the n side. Similar expressions describe the contributions of minority electrons on the p side of the diode. Taking the ratio of the two equations gives

$$i_p = Q_p \frac{D_p}{L_p^2}, \tag{20-2}$$

where D_p is the diffusion coefficient of minority holes, which measures the rate of hole diffusion, and L_p is the diffusion length, or the average distance travelled by the holes before they are lost by recombination with electrons. These two quantities are related by

$$L_p = \sqrt{D_p \tau_p}, \tag{20-3}$$

where τ_p is called the <u>minority carrier lifetime</u> for holes; it is the average time that an excess hole spends on the n side before it disappears by recombination. Combining (20-2) and (20-3) gives

$$i_p = \frac{Q_p}{\tau_p}. \tag{20-4}$$

This is the <u>charge-control equation</u> for the steady-state hole current in a forward biased pn junction diode. The total diode current is obtained by adding the corresponding term for electrons on the p side to give (20-5).

$$i_D = \frac{Q_p}{\tau_p} + \frac{Q_n}{\tau_n}. \tag{20-5}$$

The charge-control equation is just as valid for describing the diode current as the more familiar voltage-control equation

$$i_D = I_s \left(e^{\frac{v_D}{\eta V_T}} - 1 \right) \tag{20-6}$$

Equation (20-5) states that the diode current is proportional to the amount of excess charge on each side of the diode. As one might expect, if the excess carriers recombine quickly (short lifetime), then carriers must be injected at a higher rate (more current) to sustain the necessary amount of excess minority carrier charge.

The charge-control model is well suited to the analysis of the transient response of a diode. Consider a p+n diode, in which the p side is more heavily doped than the n side, so that most of the current is carried by holes, allowing the second term in (20-5) to be neglected. To include time dependence, a second term must be added, giving

$$i_D(t) = \frac{Q_p}{\tau_p} + \frac{dQ_p}{dt}. \tag{20-7}$$

This second term accounts for the rate at which the stored minority charge increases while the current is flowing. It becomes zero when steady state is reached. Equation (20-7) applies for both turn-on and turn-off of the diode.

<u>Turn-on</u>

When a voltage step is applied to a series resistor-diode circuit at time t_1, as shown in Fig. 20-1a, the diode voltage v_D at time t_1^+ is zero because some time is required to establish the minority carrier profiles; in other words, the diode acts as a capacitor. However, Ohm's Law requires that a current

$$i(t) = I_F = \frac{(V_F - v_D)}{R} \tag{20-8}$$

flow instantly in the resistor, as shown in Fig. 20-1b. (It is assumed that V_F is much greater than the steady-state value of v_D.) The diode voltage rises rapidly at first as Q_p is increased (Fig. 20-1c), then more slowly as more of the injected carriers are used to replace those lost by recombination. It eventually reaches the steady-state value V_D shown on the graph, and the second term in (20-7) becomes zero.

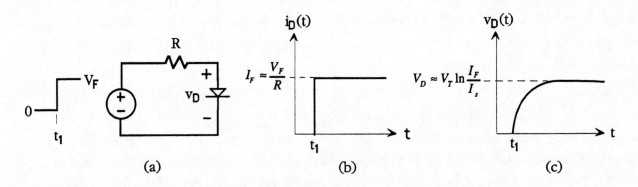

Fig. 20-1

Turn-off

At time t_2, the diode is suddenly reverse biased by changing the source voltage from V_F to a negative value V_R, as shown in Fig. 20-2a. Ohm's Law requires that the current reverse its direction immediately, as shown in Fig. 20-2b. Now,

$$i(t) = I_R = \frac{V_R - V_D}{R}. \tag{20-9}$$

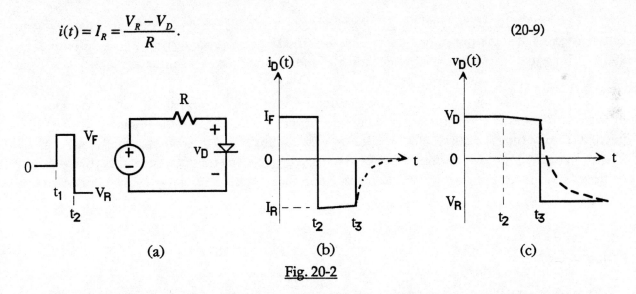

Fig. 20-2

Note in Fig. 20-2c that the voltage drop across the diode does not change much between t_2 and t_3, while the excess charge is being removed. The reverse current I_R flows in a direction opposite to the polarity of the diode voltage. This is because the diode voltage must correspond to forward bias ($v_D > 0$) as long as excess carriers remain. The reason for the slow rate of change in v_D lies in the exponential charge-voltage relationship (20-1); each tenfold decrease in the excess carrier concentration (or charge) results in a decrease in v_D of only about 60 mV. Therefore v_D can reach zero and eventually change sign only when the excess carrier concentration has been reduced by many orders of magnitude. This occurs at time t_3.

Although one might expect I_R to be very small because the diode is reverse biased, this does not happen immediately because the excess hole distribution in the n side of the diode does not disappear immediately. As long as any excess holes are there, a large reverse current I_R will flow. This flow corresponds to the removal of these excess holes. The time needed to remove them can be obtained with use of the charge-control model.

Integrating (20-7) with the initial condition $Q_p(t_2) = I_F \tau_p$ (steady-state forward bias) gives

$$Q_p(t) = \tau_p \left[I_R + (I_F - I_R) e^{\frac{-(t-t_2)}{\tau_p}} \right]. \qquad (20\text{-}10)$$

Equation (20-10) is the complete expression for the time dependence of the minority carrier charge during turn off. At some time t_3, all the excess holes have disappeared, and the current returns its steady-state reverse-bias value $-I_s$. This transition would occur almost instantaneously if there were no capacitance associated with the diode. But the capacitance of the diode's depletion layer takes time to discharge through the resistor R. This adds the exponential tail to the current waveform in Fig. 20-2b after time t_3. The time t_3 can be calculated from (20-10) by setting $Q_p = 0$ at t_3. Solving for $t_3 - t_2$ gives

$$t_3 - t_2 = \tau_p \ln\left(\frac{I_F - I_R}{-I_R}\right) \qquad (I_R < 0). \qquad (20\text{-}11)$$

The interval $t_3 - t_2$ is called the <u>diode storage time</u> t_s. The period of exponential decay due to capacitance is called the <u>transition time</u> t_t. The total diode <u>reverse recovery time</u> t_{rr} is the sum of t_s and t_t. This equation can be used to calculate the hole lifetime from time measurements.

Similar equations apply to the base-emitter diode of a bipolar transistor, with one major difference. The base region of the transistor is made very narrow, so that the minority carriers injected from the emitter can reach the collector with minimal loss by recombination in the base. The time required for their journey is the <u>base transit time</u> τ_t. The transit time, rather than the lifetime, is the governing quantity in the charge-control equations for the transistor. Obviously, $\tau_t < \tau_p$.

Procedure

1) Build the circuit in Fig. 20-3, using a 1N4007 diode or its equivalent. v_{IN} is a square wave with its frequency just low enough to allow the diode to reach steady state in both forward and reverse bias. Trigger the oscilloscope with v_{IN}, using the positive slope to observe turn-on, and the negative slope to observe turn-off. Make the oscilloscope sweep rate fast enough to allow you to measure the critical time intervals as accurately as possible. Begin with no dc offset to v_{IN} ($V_R = 0$), and record the values of v_{IN}, v_D, and $t_3 - t_2$ in your notebook.

2) Repeat the measurements with a negative offset added to v_{IN}, making $V_R = -1$ V, -2 V, -3 V, -4 V, and -5 V, while maintaining the positive amplitude at $+5$ V.

3) Repeat the measurements of parts 1 and 2, using the base-emitter diode of a bipolar transistor such as the 2N2222.

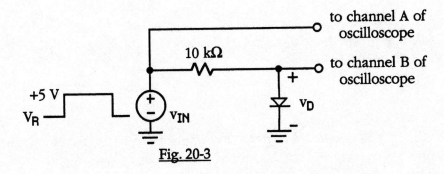

Fig. 20-3

Analysis of the Data

A) Use eqn. (20-11) to calculate the minority-carrier lifetime for each of your six sets of measurements. Does τ_p vary when the reverse-bias voltage is changed? Do you expect it to depend on the reverse bias?

B) Are your observations in part 3 consistent with your expectations? Discuss the differences between the response times of the simple diode and the transistor.

EXPERIMENT 21

DIRECT-COUPLED MULTISTAGE AMPLIFIERS

11.1 Multistage and Power Amplifiers
11.2 Two-Port Amplifier Cascade
11.3 Multistage Amplifier Biasing

Purpose

Practically all integrated-circuit amplifiers require more than one stage of amplification. For example, the op amp needs a differential input stage, an output buffer to enable it to drive low-resistance loads, and an open-loop voltage gain of the order of 10^5 or more. It is not possible to accomplish all these functions with a single stage. In this experiment you will study some of the problems encountered in designing and testing a multistage amplifier.

Introduction

Because large-value capacitors cannot be used on an integrated-circuit chip, the practice is to connect the output of one stage directly to the input of the next. In these <u>direct-coupled, or dc-coupled</u> amplifiers, the bias design is more complicated because one cannot deal with each stage independently. On the other hand, the overall voltage gain can be obtained by the Thevenin-block method that was studied in Experiment 14. It is only necessary to find the intrinsic gain and the input and output resistances for each stage, then multiply the individual gains and the loading factors.

Since it is not feasible in this experiment to work with a multistage amplifier that obeys all the integrated-circuit design rules, the compromise circuit of Fig. 21-1 will be used.

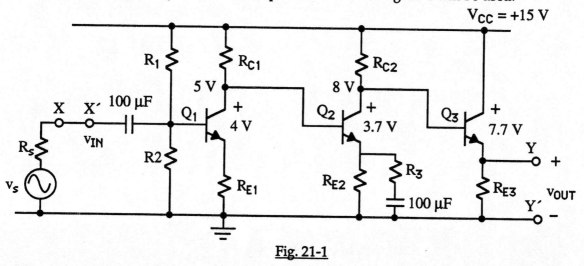

Fig. 21-1

This amplifier consists of two inverter stages, each with a voltage gain greater than unity, cascaded with an output follower. Although a large input capacitor and a large emitter bypass capacitor are used, the three stages are directly coupled, and there is no capacitor at the output.

Because of the direct coupling, the collector voltage of each stage must be higher than that of the preceding stage to prevent saturation of the transistors. Aso, the output voltage will not be zero at the operating point. You will investigate ways to overcome these liabilities in a design project.

Procedure

BEFORE COMING TO THE LABORATORY

Derive an expression for the overall voltage gain v_{out}/v_{in}, and for the input and output resistances of the circuit. This expression can be obtained by the Thevenin-block method or from an analysis of the small-signal representation of the entire circuit. The choice of method is left to you.

Choose the values of all resistors so that Q_1, Q_2, and Q_3 have approximately the indicated V_{CE} values so that certain of the node voltages are as shown, and so that the amplifier has an overall voltage gain of about +50. Aim for an intrinsic gain of about -10 for each inverter stage, then adjust to account for the input and interstage loading factors. Use any valid engineering approximations in your design.

Although an accurate determination of the high-frequency and low-frequency rolloff points of a multistage circuit is tedious, try to obtain a rough estimation, using the methods of Experiment 19.

IN THE LABORATORY

1) Energize the circuit, and verify that the bias points lie reasonably close to the target values. Record the bias values I_C and V_{CE} for each transistor in your laboratory notebook.

2) Connect a sinusoidal signal generator to the input. An attenuator, such as the one shown in Fig. 21-2, may be required on the signal generator output because its minimum peak voltage (with the amplitude control turned all the way down) may still be large enough to drive one of the transistors into saturation or into cutoff.

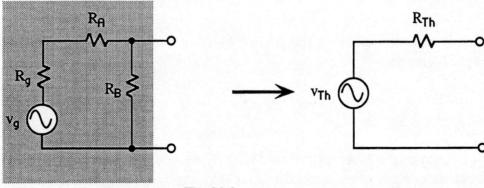

Fig. 21-2

If you use an attenuator, consider your signal source to be the Thevenin equivalent (eqn. 21-1) of the signal generator open-circuit voltage v_g, the 50-ohm internal resistance R_g of the generator, and the resistors R_A and R_B. The elements v_s and R_s on the amplifier schematic represent this source.

$$v_{Th} = v_g \left(\frac{R_B}{R_g + R_A + R_B} \right) \quad \text{and} \quad R_{Th} = (R_g + R_A) \| R_B \quad (21\text{-}1)$$

3) Measure the midband voltage gain of the amplifier, being sure that the output signal is not clipped or seriously distorted.

4) Measure the incremental input resistance of the amplifier, using the "variable resistance" method described in Experiment 14. The procedure is repeated below:

Apply an input signal in the midband region and measure the output voltage of the amplifier. Next, insert a variable resistor is series between points x and x' in the schematic, and adjust its value until the output signal is halved. Remove the resistor and measure its value with an ohmmeter. If $R_A \| R_B \ll r_{in}$, then the value of the variable resistor is approximately r_{in}. Be sure that you understand why this method works.

5) Measure the output resistance of the amplifier, again using the variable resistance method of Experiment 14:

Connect a variable resistance in series with a capacitor of appropriate value between the output node and ground (between nodes y and y'), and adjust the resistor until the output voltage is half the open-circuit value. Remove the resistor and measure its value, which will be approximately r_{out}. *Why is the capacitor needed?*

6) Now increase the value of v_s until the amplifier output is clipped, indicating that one of the transistors has reached either cutoff or saturation. Record the input and output voltages at the point where this occurs.

7) Measure the high-frequency and low-frequency rolloff points of the amplifier.

Analysis of the Data

A) Compare the observed and predicted bias currents and voltages. Try to explain any significant differences.

B) Do the same for the voltage gain. What changes might you make in the circuit to achieve a larger voltage gain?

C) Compare the observed and predicted input and output resistances of the circuit. If you were to place an emitter bypass capacitor in the first stage to increase its intrinsic gain, what would happen to the input loading factor, and to the overall gain?

D) Based on your observations in part 6 and your analysis of the transistor bias points, determine which of the transistors was the first to go out of the active region when you increased the input voltage, and whether it went into cutoff or into saturation.

E) Are the observed low-frequency and high-frequency rolloff points consistent with your predictions? Most likely you will observe significant departures from the predicted values. Can you suggest the reasons?

Design Project

Your multistage amplifier as shown in Fig. 21-1 suffers from "bias-creep". The operating-point collector voltage increases with each successive stage because the collector of each transistor must be more positive than the emitter to avoid saturation. Eventually the collector voltage will be close to V_{CC}, leaving insufficient room for signal swing. Furthermore, the output operating-point voltage is not zero. One way to overcome these defects is to add a large coupling capacitor between each stage and in series with the load. However, in keeping with the limitations of IC design, another method of <u>level shifting</u> must be used. Add another stage to your circuit to counteract the bias creep and to bring the quiescent output voltage to or near zero. No capacitors are allowed. If possible, build and test the modified circuit. Otherwise, demonstrate mathematically that your modification accomplishes its goal.

EXPERIMENT 22

CLASS AB POWER AMPLIFIER STAGE

11.6 Power Amplification Output Stages
  11.6.2 Linearly-Biased (Class A) Output Configuration
  11.6.3 Minimally-Biased (Class AB) Output Configuration

Purpose

To study one version of an output buffer stage for an amplifier that permits the delivery of sufficient power to a low-impedance load with minimal distortion of the output waveform.

Introduction

The final stage of practically all amplifiers is a buffer that can supply sufficient power to a low-impedance load. Most output stages are based on the emitter follower that was studied in Experiment 16; they provide a voltage gain that is less than unity, but a current gain approximately equal to β_F of the transistor. The simple emitter follower is a <u>Class A amplifier</u>, characterized by a highly linear voltage transfer function, but low power-transfer efficiency. The reason for the low efficiency is that bias current flows through the transistor and the load even when the signal amplitude is zero. The circuit in Fig. 22-1 is a <u>Class B amplifier</u>, which is much more efficient.

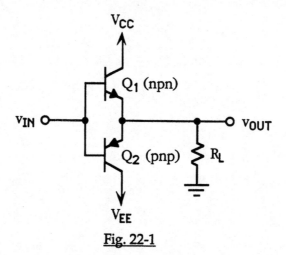

Fig. 22-1

Note that both an npn and a pnp transistor are used. When v_{IN} is zero, v_{OUT} is also zero. Both transistors are cut off, and no current flows. When v_{IN} becomes sufficiently positive to turn Q_1 on, current flows from V_{CC} through Q_1 into the load, and v_{OUT} is positive. Q_2 remains cut off because its base is more positive than its emitter. When v_{IN} becomes sufficiently negative, Q_2 is turned on and Q_1 is cut off. Current flows out of the load, through Q_2, into the V_{EE} supply. This configuration is ofter called a "<u>push-pull</u>" <u>amplifier</u>

This symmetrical amplifier is ideal from the standpoint of efficiency - current flows only when a signal is present, and none of it is wasted in the second transistor. Unfortunately, the high efficiency is gained at the expense of distortion in the output waveform. Practically no load current flows until the magnitude of the input voltage is high enough to forward bias the emitter-base junction; so the output voltage is effectively zero until the input reaches about 0.6 V, the "turn-on" voltage of the diode. The result is shown in Fig. 22-2 for the case of a sinusoidal input signal.

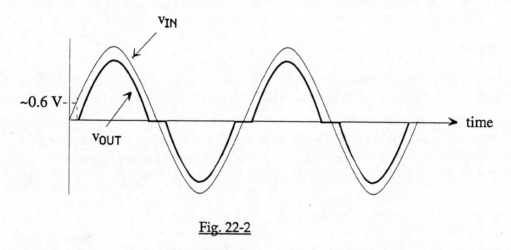

Fig. 22-2

This output waveform suffers from <u>crossover distortion</u>; it contains higher frequency components superimposed on the signal frequency. In a sound system it will appear as harmonic distortion.

There are many schemes in use to overcome the threshold voltage of the base-emitter diode, thus reducing the amount of crossover distortion. The resulting configuration is called a <u>Class AB</u> power amplifier. The one that you will study here makes use of a negative-feedback amplifier that acts as a "superdiode" (see Experiment 2). It is shown in Fig. 22-3 in its simplest form..

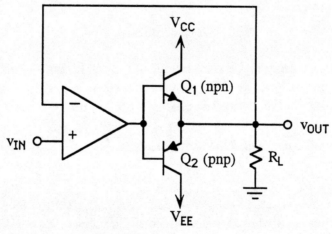

Fig. 22-3

The op amp, the base-emitter diode of one of the transistors, and the feedback loop form the superdiode. Since it is a unity-gain follower, $v_{OUT} = v_{IN}$ for all values of v_{IN}, thus eliminating the crossover distortion. What is actually happening is that an extremely small input signal is sufficient to forward-bias the diode because the input is greatly amplified by the op amp.

Additional voltage gain can be achieved by placing the usual resistor divider in the feedback loop to form a noninverting amplifier with voltage gain determined by the resistor ratio.

Note that the op amp must supply the base current for the output transistors and the current, if any, through the feedback loop. If the transistor β_F and/or the output current limit of the op amp are not high enough, the output voltage may reach a saturation level.

Procedure

1) Build the basic push-pull amplifier as shown in Fig. 22-4, using power transistors such as the TIP-31 and TIP-32 or equivalent. Set $V_{CC} = -V_{EE} = 15$ V, and $R_L = 100$ Ω with a 2-watt power rating. The signal source should be a 1-kHz sinusoid. Using the same vertical deflection setting for both scope channels, start with $v_{IN} = 0$, then increase v_{IN} until you begin to observe an output. Record the point at which the output becomes visible. Raise the amplitude of v_{IN} to about 5 V, and sketch the observed waveforms in your notebook.

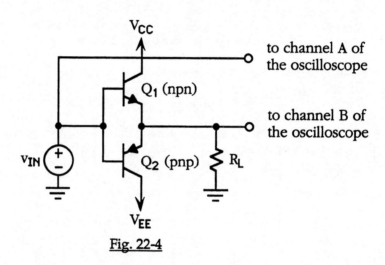

Fig. 22-4

2) Modify your circuit as shown in Fig. 22-5, with R_F around 4.7 kΩ, and R_1 a 10-kΩ potentiometer connected as a variable resistor. As before, connect v_{IN} and v_{OUT} to the two oscilloscope channels, and use the same power supplies and load resistor. Begin with R_1 set to its maximum value. Using the same waveform as in part 1, set the input amplitude to 2 V. Record the measured values of R_F and R_1 and the amplitudes of v_{IN} and v_{OUT}. Also sketch the waveforms in your notebook.

3) Repeat the measurements of part 2 for several smaller values of R_1. Continue until the output voltage shows no further increase. Record the value of R_1 at which this happens.

Analysis of the Data

A) In part 1 of the procedure you should have observed crossover distortion. Comment on your observations, especially on any deviations from expected behavior that may have occurred.

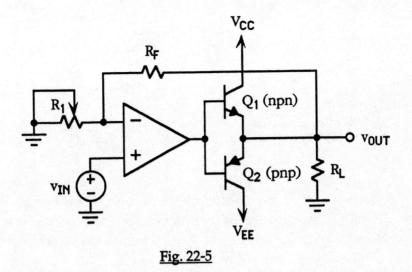

Fig. 22-5

B) Comment on the changes in output waveform that you saw in part 2. Compare your observed voltage gain with the value calculated from the <u>measured</u> values of R_1 and R_F.

C) Determine from the observations in part 3 whether the limitation on the output amplitude was caused by saturation of the output transistor or by reaching the output current limit of the op amp.

EXPERIMENT 23

FREQUENCY COMPENSATION OF AN OPERATIONAL AMPLIFIER

10.8 Feedback Loop Stability
  10.8.4 Frequency Compensation

Purpose

A feedback amplifier may become unstable and begin to oscillate when the phase shift between the input and output waveforms reaches a critical value. In this experiment you will study the conditions under which this instability occurs and a means for preventing it.

Introduction

The feedback amplifiers that you studied in Experiment 1 and elsewhere can be modeled as shown in Fig. 23-1. The amplifier block represents the op amp itself, whose open-loop voltage gain is A_0. The feedback factor β is the ratio $R_1/(R_1 + R_F)$ of the feedback resistors, which, of course, is always less than unity. The summing node Σ models the input stage of the op amp where the inverting input is subtracted from the noninverting input.

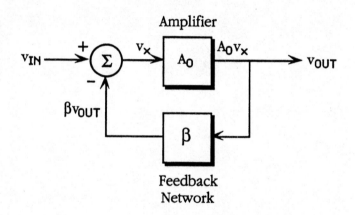

Fig. 23-1

Using this model, we can obtain the voltage transfer function with the following equations:

$$v_x = v_{IN} - \beta v_{OUT}; \qquad (23\text{-}1)$$

and

$$v_{OUT} = A_0 v_x. \qquad (23\text{-}2)$$

Combining these and solving for v_{OUT}/v_{IN} yields

$$\frac{v_{OUT}}{v_{IN}} = \frac{A_0}{1 + \beta A_0}. \qquad (23\text{-}3)$$

In midband, where the open-loop gain A_0 is extremely large, this reduces to

$$\frac{v_{OUT}}{v_{IN}} \approx \frac{1}{\beta} = \frac{R_1 + R_F}{R_1}, \tag{23-4}$$

which is the familiar gain expression for the noninverting amplifier. Since all amplifiers have a critical frequency ω_H, above which the gain begins to decrease, the approximation begins to fails at this point, where A_0 begins to decrease. Furthermore, you observed in Experiment 19 that the phase difference between the input and output signals becomes larger as the frequency increases above ω_H, until at some frequency, designated by ω_{180}, the phase difference is 180°. Fig. 23-2 shows the set of Bode plots for a hypothetical open-loop op amp with three high-frequency poles. The spacing between adjacent tick marks on the horizontal axis represents a decade of frequency. The midband gain is 120 dB (10^6). Since integrated circuits use no capacitors for bypassing or interstage coupling (see Experiment 21), there is no rolloff at low frequency. When the first pole frequency ω_1 (= ω_H) is reached, the gain begins to decrease at the rate of -20 dB/decade of frequency, and the phase shift ϕ is -45°. When each successive pole frequency is reached, the slope begins to decrease by an additional 20 dB/decade, and the phase shift becomes more negative by 90°. The phase shift becomes exactly -180° at a frequency ω_{180} that always lies between the second and third poles. *You should prove that this statement is true.* A phase shift of 180° is equivalent to a negative value of A_0.

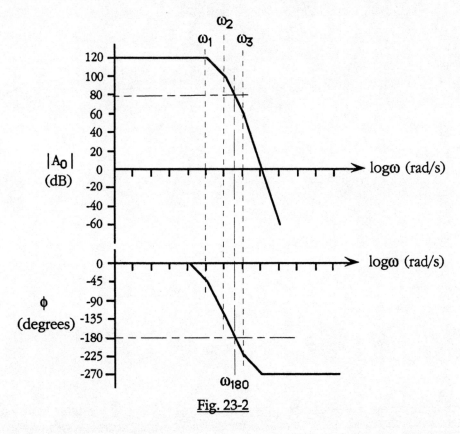

Fig. 23-2

Equation (23-3) shows that if $A_0 < 0$ and $|\beta A_0| = 1$, the gain of the feedback amplifier increases without limit. This means that an output signal is produced when there is no input; in other words, the amplifier becomes unstable, and begins to oscillate. Actually, this instability occurs when $|\beta A_0| \geq 1$. Therefore, $|\beta A_0|$ must be less than unity to prevent oscillation.

The condition $|\beta A_0| \le 1$ at ω_{180} is the stability criterion for a feedback amplifier.

In the example shown in Fig. 23-2, $|A_0| \approx 76$ dB at ω_{180}. Therefore, β must be less than -76dB ($\sim 10^{-4}$) for the stability criterion to be satisfied. Using this result in equation (23-4) leads to the conclusion that <u>any noninverting amplifier that uses this op amp and that has a voltage gain less than 10,000</u> will be unstable. This op amp is obviously not of much use as a linear amplifier.

For an op amp to be stable in the worst case of $\beta = 1$, which represents a unity-gain amplifier, the stability criterion requires that the open-loop gain A_0 itself be less than unity at ω_{180}. This can be accomplished as shown in Fig. 23-3 by adding another pole (ω_0 on the Bode plot) at a sufficiently low frequency to make A_0 begin to roll off early enough to reach unity below ω_{180}. Since ω_{180} always lies between the second and third poles, it now falls between ω_1 and ω_2, and is designated ω_{180} (new) on the plot. The dashed curves show the magnitude and phase response with the new pole added.

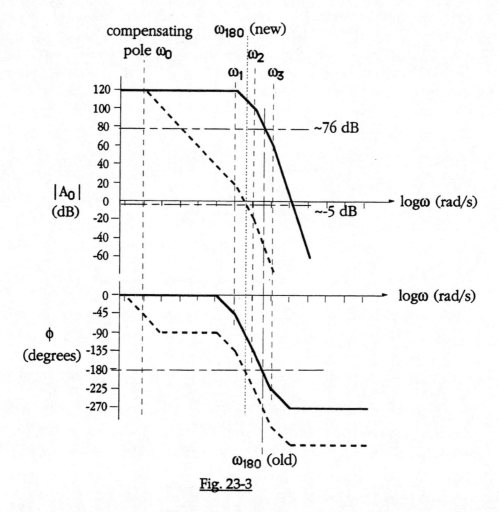

Fig. 23-3

Now that the op amp has been <u>frequency compensated</u>, it will be stable in any negative-feedback configuration because $|A_0| < 0$ at ω_0. The cure for instability seems to be a drastic one, however. The high-frequency rolloff point had to be reduced by five orders of magnitude! The LM741 op amp, for example, which is compensated internally, has $f_0 < 10$ Hz. Although this is a very low frequency, one must remember that an open-loop op amp is never used as a linear amplifier. If the LM741 is used in a feedback configuration to give a voltage gain of 100, for example, the bandwidth will be around 10^5 Hz. The reason for this can be seen in Fig. 23-3, where a horizontal line at |gain| = 40 dB (= 100) intersects the open-loop gain curve about four decades above ω_0.

The new pole can be created in many ways. Perhaps the simplest way is with an external RC network in which $\omega_0 = 1/RC$. Two possible ways are shown in Fig. 23-4.

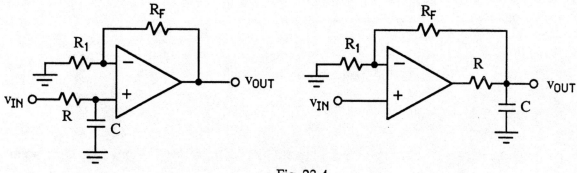

Fig. 23-4

This approach has the disadvantage of needing large values of R and C to create a pole near 10 Hz. A better way is to modify the op amp itself. Some op amps, such as the LM741, have a capacitor built into the IC chip, as shown in Fig. 23-5.

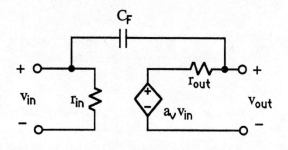

Fig. 23-5

The schematic shows the Thevenin equivalent of the high-gain stage of the op amp, for which the intrinsic gain a_v is around 10^6. The capacitor forms a feedback path from output to input, and is therefore subject to the Miller effect, just as is the collector-base capacitance C_μ inside a transistor. Because of the extremely high gain, the Miller effect makes a 30-pF capacitor acts as if it were around 30 μF. The pole associated with this large effective capacitance is therefore at a low frequency.

Some op amps, such as the LM301 and the LM748, permit the designer to set the compensating pole frequency by providing two terminals to which an external C_F of the order of 10 pF can be connected.

Because the compensating pole is at such a low frequency, the op amp acts essentially as a single-time-constant amplifier, with a rolloff rate of -20 dB/decade over practically all of its usable frequency range. Its open-loop gain can therefore be described by

$$A(j\omega) = \frac{A_0}{1 + j\omega/\omega_0},$$ (23-5)

where ω_0 is the frequency of the compensating pole, and A_0 is th emidband gain. Substituting $A(j\omega)$ for A in (23-3) gives the gain with feedback.

$$A_{fb} = \frac{V_{OUT}}{V_{IN}} = \left(\frac{A_0}{1+\beta A_0}\right)\left(\frac{1}{1+j\omega/[\omega_0(1+\beta A_0)]}\right).$$ (23-6)

The feedback amplifier now has its dominant pole ω_{fb} at

$$\omega_{fb} = \omega_0(1+\beta A_0).$$ (23-7)

Feedback therefore increases the bandwidth of the amplifier by the factor $(1+\beta A_0)$. This situation is shown in Fig. 23-6 for two values of A_{fb} (+20 dB and +60 dB).

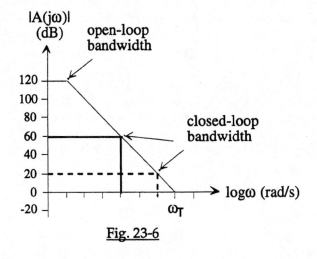

Fig. 23-6

Rearranging (23-7) and multiplying both sides by A gives

$$\frac{A_0}{1+\beta A_0}\omega_{fb} = A_0\omega_0.$$ (23-8)

The left-hand side is the product of the closed-loop gain A_{fb} and the closed-loop bandwidth. The right side is the product of the open-loop gain and the open-loop bandwidth, which is constant for a given op amp. Therefore, the gain-bandwidth product of a feedback amplifier is constant. This can be seen in Fig. 23-6. Reducing the gain of a feedback amplifier from 60 dB to 20 dB, a factor of 100, increases the bandwidth by two decades, or a factor of 100. A figure of merit for the op amp is its <u>unity-gain frequency</u> ω_T, or the frequency where $A(j\omega) = 1$, or 0 dB. The gain-bandwidth product of any feedback amplifier made with the op amp equals ω_T.

Procedure

1) Build the inverting amplifier shown in Fig. 23-7, using an LM301 op amp or a similar one that allows external compensation, and a 10-kΩ potentiometer in the feedback loop. The 8-pin DIP version of the op amp most likely uses pins 1 and 8 for connecting the external compensating capacitor (see Appendix C); but to be sure, you should consult the manufacturer's data sheet for your op amp.

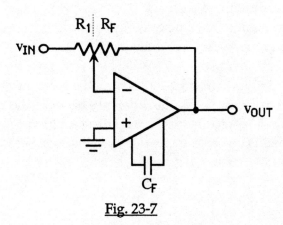

Fig. 23-7

Begin with C_F close to 30 pF and the potentiometer adjusted to give a voltage gain as close to unity as you can achieve. Use a constant-amplitude sinusoidal input signal set to give an output amplitude at low frequency (midband) of no more than 0.1 V. This is done to insure that the amplifier will not become slew-rate-limited at higher frequencies.

2) Observe v_{OUT} while you increase the signal frequency, and record the frequency f_H at which v_{OUT} is reduced from the midband value by approximately $1/\sqrt{2}$, or 0.707. You should display both v_{IN} and v_{OUT} simultaneously to insure that v_{IN} does not change with changing frequency.

3) Repeat the measurement of f_H with the same 30-pF capacitor, but with the potentiometer set to give voltage gains of about -10 and -100. Again observe both v_{IN} and v_{OUT} simultaneously on the scope to be sure of your voltage gain.

4) Reset the potentiometer to give a gain of -100. Then replace the 30-pF capacitor with a 5-pF capacitor, and observe v_{IN} and v_{OUT}. If the amplifier behaves as expected at low frequency, measure and record f_H as you did in part 1. Return to a low frequency and begin to adjust the potentiomter, reducing the gain in small steps. At some point oscillations should occur, and should persist even when v_{IN} is reduced to zero. Record the frequency and amplitude of the oscillations and the feedback factor R_F/R_1 at which oscillation accurred. If you do not observe oscillations, you may have to try a smaller compensating capacitor.

5) Use a compensating capacitor of about 10 pF, and try different feedback factors to obtain oscillation again. Record the new oscillation frequency and the feedback factor at which it occurred.

Analysis of the Data

A) Find the gain-bandwidth product and the unity-gain frequency f_T from the data of parts 2 and 3. The 30-pF capacitor should have placed the compensating pole at a sufficiently low frequency to make the op amp stable even at unity gain.

B) Find the gain-bandwidth product and f_T from the measurement in part 4. The difference between the two f_T values is approximately the difference between the compensating poles introduced by the 30-pF and 5-pF capacitors. A capacitor near 5 pF is insufficient to reduce A_0 below 0 dB at f_{180}. You should be able to verify this by showing from your data in parts 4 and 5 that f_T is greater than the oscillation frequency, which is f_{180}.

EXPERIMENT 24

COMPARISON OF PASSIVE AND ACTIVE FILTERS

1. *Review of Linear Circuit Theory*
    1.9 *Single-Time-Constant Resistor-Capacitor Circuits*
       1.9.2 *RC Circuit Response in the Sinusoidal Steady State*
13. *Active Filters and Oscillators*
    13.1 *A Simple First-Order Active Filter*

Purpose

To compare the characteristics of passive and active versions of a first-order low-pass filter that have the same rolloff frequency.

Introduction

Both of the circuits shown in Fig. 24-1 are low-pass filters.

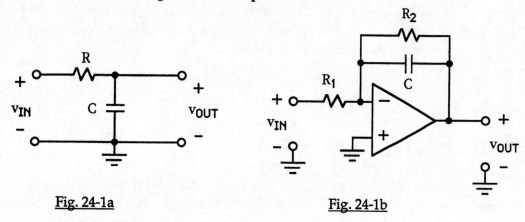

Fig. 24-1a   Fig. 24-1b

Using voltage division in the passive network in Fig. 24-1a gives

$$\frac{\mathbf{V}_{out}}{\mathbf{V}_{in}} = \frac{1}{1+j\omega RC} \qquad (24\text{-}1)$$

The boldface symbols indicate that the voltages are phasor quantities. The corresponding function for the active version in Fig. 24-1b is derived from the gain expression for the inverting amplifier

$$\frac{\mathbf{V}_{out}}{\mathbf{V}_{in}} = -\frac{\mathbf{Z}_{feedback}}{R_1}, \qquad (24\text{-}2)$$

where

$$Z_{feedback} = R_2 \left\| \frac{1}{j\omega C} \right. . \qquad (24\text{-}3)$$

This results in

$$\frac{V_{out}}{V_{in}} = -\frac{R_2}{R_1}\left(\frac{1}{1+j\omega R_2 C}\right). \qquad (24\text{-}4)$$

When R in the passive circuit equals R_2 in the active circuit, and the capacitance is the same in both circuits, the bandwidth ω_H, or the frequency at which the voltage ratio is reduced by 3 dB, will be 1/RC for both.

The active filter is superior to the passive filter in two important ways:

a) the magnitude of the transfer function of the active version at low frequency ($\omega \rightarrow 0$) equals R_2/R_1, which can be considerably greater then unity, whereas the gain of the passive network is always less than unity.

b) A low-resistance load has no effect on the output of the active filter because the feedback amplifier has an extremely low output resistance. On the other hand, both the magnitude of the transfer function and the rolloff frequency of the passive filter are changed when a load is connected across its output terminals.

Procedure

BEFORE COMING TO THE LABORATORY

Build the two filters shown in Fig. 24-1 on the same breadboard, using the following component values: passive filter - C = 10 nF, R = 2200 Ω; active filter - C = 10 nF, R_1 = 220 Ω, R_2 = 2200 Ω. Calculate the expected rolloff frequency and the voltage gain for each.

IN THE LABORATORY

1) Excite the input ports of both filters with a sinusoidal signal of 0.5-V amplitude, using ±15-V power supplies for the op amp. Use the oscilloscope with no load resistor connected at the output node to measure the amplitude of the output voltage of each filter at several frequencies. Begin around 1 Hz, and increase the frequency to the point where the output is reduced to no more than 10 percent of its low-frequency value. Record enough points near the rolloff frequency to allow you to make an accurate plot of amplitude vs frequency.

2) Repeat the measurements of part 1 with a 220-Ω load resistor connected to the output port of each filter.

Analysis of the Data

A) Make Bode magnitude plots for each filter, with and without the load resistor. Recall that a Bode plot displays the gain magnitude in dB ($20\log|v_{OUT}/v_{IN}|$) vs $\log\omega$. Find the midband gain, the rolloff frequency ω_H, and the rolloff rate above ω_H, in dB/decade of frequency, for the four cases. Recall that ω_H is the frequency where the output voltage is 3 dB below the midband value, or $1/\sqrt{2}$ times the midband value.

B) Compare the two filters with no load and with the load. Do the results agree with predictions? Your measured values of gain and bandwidth may be different from your calculated values. If so, discuss the possible reasons for the differences.

C) What are your conclusions concerning the relative merits and limitations of the active and passive versions of the low-pass filter?

EXPERIMENT 25

SECOND-ORDER ACTIVE LOW-PASS FILTER

13.3 Second-Order Filter Responses
    13.3.1 The Biquad Fiter Function
    13.3.2 Second-order Active Low-Pass Filter

Purpose

In Experiment 24 you studied a first-order active filter, which contains one reactive element, and which has a rolloff rate of -20 dB/decade outside the passband. In this experiment you will make a comparison with a second-order filter, and investigate frequency scaling of a filter.

Introduction

Figure 25-1 shows an active filter with two reactive elements, and therefore two poles. Its voltage-transfer function, derived by nodal analysis at the node between the two resistors, is

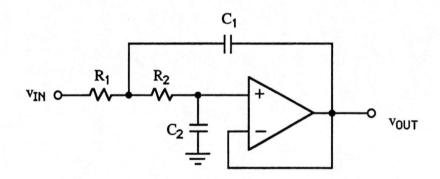

Fig. 25-1

$$\frac{\mathbf{V}_{out}}{\mathbf{V}_{in}} = \frac{1}{Z_1 Z_2 + Z_1(R_1 + R_2) + R_1 R_2}. \tag{25-1}$$

The bold-face symbols are the phasor representations of the voltages. Substituting the expression $1/j\omega C$ for the impedances gives

$$\frac{\mathbf{V}_{out}}{\mathbf{V}_{in}} = \frac{1}{s^2(R_1 R_2 C_1 C_2) + sC_2(R_1 + R_2) + 1}, \text{ where } s = j\omega. \tag{25-2}$$

By making the substitutions

$$\omega_0 = \frac{1}{\sqrt{R_1 R_2 C_1 C_2}} \text{ and } Q = \frac{\omega_0 R_1 R_2 C_1 C_2}{C_2(R_1 + R_2)} = \sqrt{\frac{C_1}{C_2}}\left(\frac{\sqrt{R_1/R_2}}{1 + R_1/R_2}\right), \tag{25-3}$$

(25-2) can be written as

$$\frac{V_{out}}{V_{in}} = \frac{\omega_0^2}{s^2 + \frac{s\omega_0}{Q} + \omega_0^2}. \qquad (25\text{-}4)$$

The measured voltage-transfer function is the magnitude of (25-4).

$$\left|\frac{v_{OUT}}{v_{IN}}\right| = \frac{\omega_0^2}{\sqrt{\omega^4 + \omega^2\omega_0^2\left(\frac{1}{Q^2} - 2\right) + \omega_0^4}}. \qquad (25\text{-}5)$$

The quantity ω_0 is approximately the rolloff frequency, and Q is a measure of the magnitude of the function at $\omega = \omega_0$. Note that the function approaches unity for $\omega \ll \omega_0$, zero for $\omega \gg \omega_0$, and increases without limit at $\omega = \omega_0$ when Q is very large. The circuit is obviously a low-pass filter, but it may have a peak in its response at or near $\omega = \omega_0$. Fig. 24-2 shows a graph of the filter function (25-5) for the case of $\omega_0 = 10^4$ and for five values of Q. You can prove, using (25-5), that the gain at ω_0 equals Q.

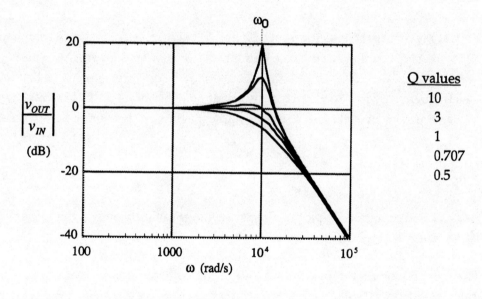

Fig. 25-2

The filter is <u>optimally flat</u> when $Q = 1/\sqrt{2} = 0.707$; substitution of this value in (25-5) shows that the gain is -3 dB at $\omega = \omega_0$. In this case there is no peaking of the function before rolloff, and $\omega_H = \omega_0$. Since the resistor and capacitor values are independent, it is possible to choose different sets of component values that maintain constant ω_0 while giving different Q values, or vice versa.

The rolloff frequency can be changed without affecting the shape of the Bode plot (no change in Q) by changing the capacitor and/or resistor values without changing their ratios, This design technique is called <u>frequency-scaling</u>. Also, the Q can be changed without changing ω_0 by changing the capacitor and/or resistor ratios while maintaining their products constant.

Procedure

BEFORE COMING TO THE LABORATORY

Calculate a set of reasonable resistor and capacitor values for the circuit of Fig. 25-1 that give $f_0 \approx 10$ kHz and $Q \approx 0.1$.

IN THE LABORATORY

1) Build the low-pass filter with your selected components. Remember that you may have to measure your component values to achieve the predicted characteristics.

2) Using a sinusoidal input signal of 1-V amplitude, measure the output voltage as a function of frequency, beginning below 100 Hz, and increasing the frequency until the output is reduced by at least a factor of 20 below the midband value. You won't need many data points between 100 Hz and several kHz, but they should be spaced about equally on the logarithmic frequency scale. Be sure to record enough points near ω_0 to allow you to generate an accurate Bode plot.

3) Repeat the measurements for a set of R and C values that maintain the same ω_0 while increasing Q to about $1/\sqrt{2}$. Repeat the calculations and measurements for $Q \approx 3$ and $Q \approx 10$.

4) Use <u>frequency scaling</u> to change ω_0 by about an order of magnitude while maintaining $Q = 1/\sqrt{2}$. This requires a new choice of resistor and capacitor values derived from (25-3).

Analysis of the Data

A) Make a Bode magnitude plot of all four sets of measurements on the same piece of log-log graph paper (20log $|v_{OUT}/v_{IN}|$ vs log ω).

B) Compare the observed maximum value of the voltage-transfer functions (near ω_0) with the values calculated from (25-5).

C) Calculate the rolloff rate of the function (in dB/decade of frequency), and compare it with what you expect for a 2nd-order filter.

Design Project

The filter shown in Fig. 25-1 has a voltage gain of unity in midband because the op amp functions as a unity-gain amplifier, with 100 percent feedback. The midband gain can be increased by incorporating a pair of resistors in the feedback loop to reduce the amount of feedback, as is done in the conventional non-inverting amplifier. Modify your filter in this way, using a variable resistor as one of the feedback components. For a fixed input signal, observe the amplitude and shape of the output waveform as the gain is increased, and comment on your observations.

You can compare your observations with theory by deriving the voltage transfer function for your new filter. The method of derivation is the same nodal analysis that was used in Fig. 25-1, except that in this case the voltage at the inverting terminal is not v_{OUT}, but a fraction of v_{OUT} determined by the ratio of the negative feedback resistor pair. The new function can be written in the form of (25-4) with Q replaced by an expression that can be interpreted as an "effective Q".

Use the new voltage-ransfer function to predict the range of voltage gains over which the filter is stable.

EXPERIMENT 26

OSCILLATORS

13.7 Oscillators
    13.7.1 Wien-Bridge Oscillator
    13.7.5 Schmitt Trigger Oscillator

Purpose

To study circuits that are inherently unstable because of the presence of positive feedback. They generate a time-varying output signal with no need for an input signal.

Introduction

We saw in Experiment 1 that negative feedback in an amplifier forces the amplifier to operate in its linear range, giving a stable output that is proportional to the input, or inputs. When positive feedback is present to oppose the negative feedback, the amplifier may reach an unstable state, or, stated another way, its gain may become "infinite", resulting in an output signal for a zero input signal. The amplifier becomes an <u>oscillator</u>. You will study two kinds of oscillator in this experiment.

I. Wien-Bridge Oscillator

The Wien-bridge oscillator shown in Fig. 26-1 uses both negative and positive feedback. We will show that it generates an output signal even though both input terminals are grounded.

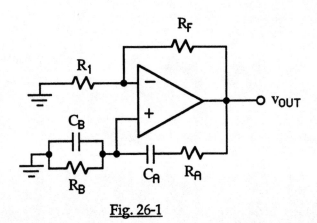

Fig. 26-1

We begin by assuming that there will be a steady-state sinusoidal output signal, and further, that the op amp remains in its linear region because there is negative feedback. This allows use of the approximation $v_+ \approx v_-$. **This kind of heuristic reasoning is frequently used in engineering and scientific analysis. However, the conclusion to which it leads must always be tested to see** whether it is a reasonable one that is consistent with the assumption. If it is not, then the assumption must be discarded.

Using voltage division in the frequency domain (phasor notation), we obtain

$$\mathbf{V}_+ = \mathbf{V}_- = \mathbf{V}_{OUT}\left[\frac{R_1}{R_1 + R_F}\right] = \mathbf{V}_{OUT}\left[\frac{\frac{1}{j\omega C_B}\|R_B}{\frac{1}{j\omega C_B}\|R_B + \left(\frac{1}{j\omega C_A} + R_A\right)}\right]. \quad (26\text{-}1)$$

Equating the two expressions in square brackets and rearranging gives

$$\frac{R_1}{R_1 + R_F} = \frac{j\omega R_B C_A}{1 + j\omega(R_B C_A + R_A C_A + R_B C_B) - \omega^2 R_A R_B C_A C_B}. \quad (26\text{-}2)$$

Since a real quantity and a complex quantity can be equal only if the imaginary part is zero, it is required that

$$\omega^2 R_A R_B C_A C_B = 1, \quad \text{or} \quad \omega = \frac{1}{\sqrt{R_A R_B C_A C_B}}. \quad (26\text{-}3)$$

The assumption therefore makes sense under a restricted set of conditions. There is a sinusoidal output at only one frequency, given by (26-3). But this places a constraint on the choice of resistor and capacitor values that results from equating the real parts of the bracketed expressions in (26-1).

$$\frac{R_1}{R_1 + R_F} = \frac{R_B C_A}{R_B C_A + R_A C_A + R_B C_B}. \quad (26\text{-}4)$$

If we choose $R_A = R_B$ and $C_A = C_B$, then

$$\frac{R_1}{R_1 + R_F} = \frac{1}{3}, \quad \text{or } R_F = 2R_1. \text{ Also, } \omega = \frac{1}{R_A C_A}. \quad (26\text{-}5)$$

In practice, R_F should be slightly greater then $2R_1$ to overcome losses in the circuit that would cause the oscillations to die out after a few cycles.

The simple Wien-bridge oscilllator in Fig. 26-1 gives a distorted output unless the output voltage is prevented from reaching its saturation levels. For this reason some sort of control network is usually added to limit the output. One such arrangement is shown in Fig. 26-2.

The circuit inside the dashed lines is the same as in Fig. 26-1. The four additional resistors and two diodes limit the output. When the output is positive, diode D_2 becomes forward biased, and clamps the output voltage to a value determined by R_2, R_3, and V_{EE}. The value of v_{OUT} can be calculated from the following equations:

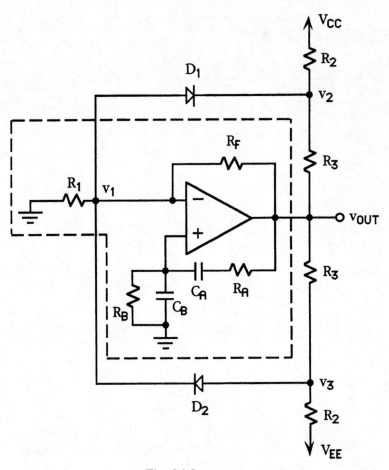

Fig. 26-2

$$v_3 \approx v_1 + 0.7; \tag{26-6}$$

$$v_1 = v_{OUT}\left(\frac{R_1}{R_1 + R_F}\right) \approx 1/3\, v_{OUT}; \tag{26-7}$$

$$\frac{v_{OUT} - v_3}{R_3} \approx \frac{v_3 - V_{EE}}{R_2} \quad \text{(neglecting the diode current);} \tag{26-8}$$

The result is

$$v_{OUT} = \frac{0.7\left(1 + \dfrac{R_3}{R_2}\right) - V_{EE}\dfrac{R_3}{R_2}}{1 - \left(\dfrac{R_1}{R_1 + R_F}\right)\left(1 + \dfrac{R_3}{R_2}\right)} = \frac{0.7\left(1 + \dfrac{R_3}{R_2}\right) - V_{EE}\dfrac{R_3}{R_2}}{\dfrac{2}{3} - \dfrac{1}{3}\dfrac{R_3}{R_2}}. \tag{26-9}$$

II. Astable Multivibrator, or Schmitt-Trigger Clock

When a negative feedback path consisting of a resistor R_x and a capacitor C_x is added to the Schmitt trigger that you studied in Experiment 2, the circuit has no stable state; the output will continually switch between its two extremes at a rate determined by the time constant $R_x C_x$. The circuit is shown in Fig. 26-3. In this case the two feedback mechanisms do not balance each other, as they did in the Wien-bridge oscillator. The positive feedback dominates, forcing the output to either its most positive or most negative limit.

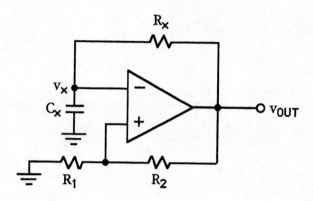

Fig. 26-3

Immediately after a transition of the output to either its positive extreme (V_{CC}) or its negative extreme (V_{EE}), the $R_x C_x$ network will begin an exponential transition; the capacitor will begin to charge or discharge, depending on its previous state, with its voltage approaching the new value of v_{OUT}. When the capacitor voltage v_- passes the value of v_+, which is determined by R_1 and R_2, the op amp output will suddenly switch to its opposite extreme. The capacitor voltage will then begin to change in the opposite direction until switching occurs again. The process will be repeated indefinitely, giving a square-wave output waveform without the need for an input voltage source.

Consider the special case where $R_1 = R_2$. The reference voltage v_+ will therefore be either $V_{CC}/2$ or $V_{EE}/2$. Suppose that at $t = 0$ the output voltage is V_{EE}, and the capacitor voltage v_x has just fallen below $V_{EE}/2$. The output will switch from V_{EE} to V_{CC} because $v_+ - v_-$ has just become positive. The capacitor voltage begins to increase, and is given by

$$v_x(t) = V_{CC}\left(1 - e^{\frac{-t}{R_x C_x}}\right) + \frac{V_{EE}}{2} e^{\frac{-t}{R_x C_x}}. \tag{26-10}$$

This is derived from the differential equation of the $R_x C_x$ network. Substitution of $t = 0$ shows that (26-10) indeed satisfies the initial condition $v_x(0) = V_{EE}/2$ and the final condition $v_x(\infty) = V_{CC}$. The capacitor voltage begins to increase toward V_{CC}, reaching $V_{CC}/2$ at time t_1. At this point $v_+ - v_-$ changes sign, and v_x begins to decrease, now governed by the equation

$$v_x(t) = V_{EE}\left(1-e^{\frac{-(t-t_1)}{R_xC_x}}\right) + \frac{V_{CC}}{2}e^{\frac{-(t-t_1)}{R_xC_x}}. \tag{26-11}$$

At time t_2, v_x reaches $V_{EE}/2$. Solving (26-11) for this condition gives

$$t_2 - t_1 = R_xC_x \ln\left(\frac{V_{EE}-V_{CC}/2}{V_{EE}/2}\right). \tag{26-12}$$

Finally, for symmetrical power supplies ($V_{EE} = -V_{CC}$),

$$t_2 - t_1 = R_xC_x \ln 3 = 1.1 R_xC_x. \tag{26-13}$$

The period of the square wave is twice this interval, and is not a function of R_1 and R_2.

Procedure

I. Wien-Bridge Oscillator

1) Build the oscillator in Fig. 26-1, choosing $R_A = R_B$ and $C_A = C_B$, and values designed to give an oscillation frequency near 1 kHz. Remember to use measured component values, not nominal values. Use $R_F = 1\ k\Omega$ and a 10-kΩ potentiometer connected as a variable resistor for R_1. Begin with R_1 at its maximum, and decrease it until oscillations start. Measure the value of R_1 at which oscillations began. *Is your output signal badly distorted?* If it is, build the improved oscillator in Fig. 26-2 if you have time. Choosing $R_3/R_2 = 1/3$ and $V_{CC} = -V_{EE} = 15$ V will limit the output to around 10 V.

2) Try other RC combinations to change the oscillation frequency. Use variable resistors to avoid frequent changing of components. Investigate the frequency range over which the oscillator can operate.

II. Astable Multivibrator

1) Build the circuit in Fig. 26-3, using a 1-μF capacitor, R_x in the range of 22 kΩ to 47 kΩ, and $R_1 = R_2$ in the range of 1 kΩ to 10 kΩ. Observe $v_{OUT}(t)$ and $v_-(t)$ simultaneously with the oscilloscope. Be sure to use the 10x scope probe when observing v_-. Compare the positive and negative peak values of v_+ and v_{OUT}, and *explain your observations*.

2) Repeat the measurements with a different capacitor value, or with a variable resistor used for R_x.

Analysis of the Data

Astable Multivibrator

Compare the observed periods of the square-wave output with equation (26-13). Predict how the output waveform will change if $R_1 \neq R_2$.

Is the output waveform truly a square wave when you observe it with the oscilloscope set at a high sweep? If not, how do you explain its shape?

EXPERIMENT 27

NMOS AND CMOS LOGIC GATES

14.2 CMOS Logic Family
    14.2.1 CMOS Inverter Transfer Characteristic
    14.2.3 CMOS Logic Gates
14.3 NMOS Logic Family
    14.3.1 NMOS Inverter with Enhancement Load

Purpose

In Experiment 12 you measured the voltage-transfer characteristic of an MOS inverter with a resistor as the pullup element. Although that configuration was useful for studying the MOSFET's characteristics, it is never used in logic circuits. In digital MOSFET IC chips the pullup element is always an MOS, partly because it occupies much less chip area than does a resistor, and partly because, if properly chosen, it provides better performance than does a resistor pullup. In this experiment you will study all-MOS inverters and some simple logic gates.

Introduction

The inverter is the basic element of not only most linear amplifiers, but of most digital logic circuits as well. In digital circuits its purpose is to act as a switch, whose output voltage is either the minimum or the maximum possible value when its input voltage is at maximum or minimum, respectively. Figure 27-1 shows the voltage transfer function of an ideal logic inverter.

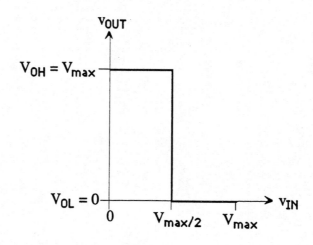

Fig. 27-1

The output voltage that corresponds to logic level "1", defined as V_{OH}, should be the maximum value that can be achieved in the circuit; the output that corresponds to logic level "0", defined as V_{OL}, should be zero. The steep slope in the transition region is needed so that the output is unequivocally high or low, and never at an intermediate value, where it cannot be interpreted as either a logic "1" or a logic "0". The ideal logic inverter is therefore an amplifier with infinite voltage gain.

In this experiment you will study the two types of MOSFET inverter shown in Fig. 27-2.. Note the directions of the arrows on the transistor symbols. The NMOS inverter uses two enhancement-mode NMOS devices, and the CMOS (complementary MOS) uses an enhancement-mode PMOS as the pullup device. In the CMOS configuration the input is applied to both gates. On integrated circuit chips the substrate of each NMOS is connected to the most negative point in the circuit, and the substrate of each PMOS is connected to the most positive point. This insures that no junction will ever become forward biased with respect to its substrate.

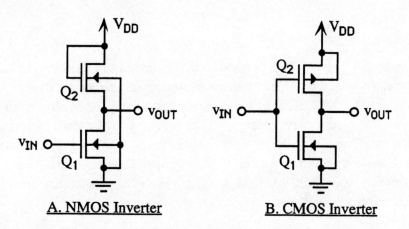

A. NMOS Inverter   B. CMOS Inverter

Fig. 27-2

A. NMOS Inverter

In this configuration, both the input element and the pullup element are enhancement-mode NMOS transistors. Because $v_{GS2} = v_{DS2}$ and $V_{TR} > 0$, Q_2 always operates in the constant-current region. Therefore, when the input voltage is low, and Q_1 is cut off, making the current essentially zero, there is still a voltage drop of V_{TR} across Q_2. At this point the output is at logic level "1", which is less than V_{DD} by an amount equal to the threshold voltage. Current begins to flow, and the output voltage begins to drop, when the input voltage exceeds V_{TR1}. The input level

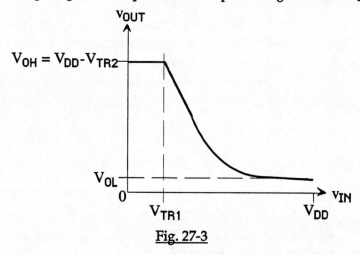

Fig. 27-3

eventually becomes high enough to force Q_1 into the triode region. Above this point the output voltage decreases slowly with increasing v_{IN}, reaching logic level "0". The voltage transfer function thus looks like Fig. 27-3.

The slope, or voltage gain, in the transition region is given by

$$gain = -\sqrt{\frac{K_1}{K_2}} = -\sqrt{\frac{W_1/L_1}{W_2/L_2}} \qquad (27\text{-}1)$$

when the MOSFETs have the same threshold voltage, as is usually the case. Recall that

$$K = \mu C_{ox} \frac{W}{L} = K' \frac{W}{L}, \qquad (27\text{-}2)$$

where μ is the mobility of the charge carriers in the channel and C_{ox} is the capacitance per unit area of the gate insulator. W and L are the channel width and length, respectively, of the MOSFET. On an enhancement-mode NMOS IC chip, the transistors are identical in all respects except for W and L, which are specified by the circuit designer. The ratio of the two W/L factors is called the aspect ratio of the inverter. The gain of this NMOS inverter is actually less than the value predicted by (27-1) because of the body effect (Experiment 9), which makes the effective threshold voltage of Q_2 higher than that of Q_1, especially when the output is high, because at this point the voltage drop between the source and the substrate of Q_2 is at its maximum.

B. CMOS Inverter

The CMOS inverter (Fig. 27-2B) operates symmetrically because the input affects both transistors directly. When the input is low, Q_1 is cut off; since no current flows, the output is at V_{OH}. At the same time Q_2 has the maximum value of $|v_{GS}|$, and is therefore in the triode region. When the input goes high, the transistors exchange roles; Q_2 is in the triode region and Q_2 is cut off. The K factors of the NMOS and PMOS are equal to insure symmetry. This requires that W_2/L_2 be approximately twice W_1/L_1 to compensate for the lower mobility of positive holes, which are the charge carriers in the PMOS. The voltage gain is very high, limited only by the incremental drain resistance r_0 of the MOSFETs. There is no body effect because both sources are connected to their respective substrates. CMOS is more difficult to manufacture than NMOS, but its improved performance far outweighs the disadvantage of increased complexity. Consequently, practically all modern MOSFET logic chips use the CMOS configuration. A typical CMOS voltage transfer function is shown in Fig. 27-4. The CMOS inverter is obviously much closer to the ideal than is the NMOS.

<u>Procedure</u>

I. NMOS Inverter

1) Use two of the NMOS devices on the CD4007 chip shown schematically in Fig. 27-5 to build the inverter of Fig. 27-2A with $V_{DD} = 5$ V. Observe that the substrates of all the NMOS

elements are connected together within the chip, but only the lower left device has an on-chip connection between its source and the substrate. This element can be used for Q_1, but not for Q_2, which must not have its source connected to the substrate.

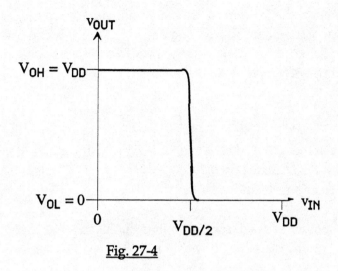

Fig. 27-4

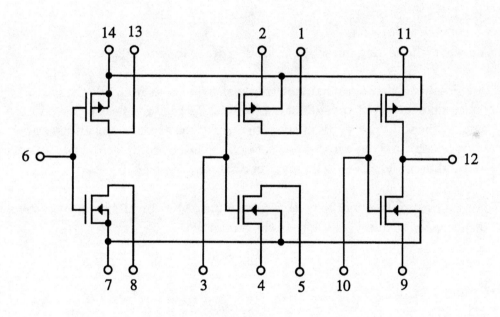

Fig. 27-5

2) Set the oscilloscope to x-y mode with full-scale deflection of 5 V on both axes. Adjust the horizontal and vertical position controls to center the spot on the graticule.

CAUTION! Make sure that the intensity control is turned low whenever a single spot is displayed. If the intensity is too high, especially if you see a halo around the spot, the screen may become permanently damaged.

3) Adjust the amplitude and dc-offset controls on your signal generator so that it produces a zero-to 5-V triangular or sinusoidal waveform at about 200 Hz. Use this waveform as the input

signal to your inverter and to the x-input of the scope. Connect the inverter output to the y-input. Sketch the observed waveform in your notebook as accurately as you can, concentrating on the values of V_{OH}, V_{OL}, and the shape and position of the transition region between them.

4) Now connect the inverter input to a dc source. Place a milliammeter in series with the transistors to measure the drain current, and use a digital voltmeter to measure v_{IN} and v_{OUT} alternately. Record readings of v_{OUT} and i_D for v_{IN} = 0, +1 V, several values in the region where v_{OUT} changes rapidly, +4 V, and +5 V.

II. CMOS Inverter

1) Use one of the NMOS device and one of the PMOS devices to build the inverter of Fig. 27-2B. The leftmost PMOS may be used as Q_2. If another PMOS is used, there must be an external connection between its source and substrate.

2) Make the same series of measurements on this inverter that you did for the NMOS configuration, and record the same information.

Analysis of the Data

A) Answer the following questions for both types of inverter:

- How closely do your observed voltage transfer functions resemble the ideal?
- What are the threshold voltages of the two MOSFETs used in each inverter?
- Calculate the static power dissipation, defined as the average power dissipated in the logic-high and logic-low states, for both inverters; compare the two. Under what conditions does the CMOS inverter dissipate power?

B) Using the expression for the voltage gain of the CMOS inverter, can you estimate the value of r_0, the incremental drain resistance of the transistors?

Design Project

Design and build a 2-input CMOS NOR gate and a 2-input CMOS NAND gate, using the CD 4007 chip. For the NOR gate, measure the voltage transfer characteristic between the output and each of the inputs separately while holding the other input at logic "0". Make the same measurements for the NAND gate, holding the other input at logic "1".

EXPERIMENT 28

PROPAGATION DELAY TIME OF A CMOS INVERTER

14.2.2 Dynamic Behavior of CMOS Logic Gates

Purpose

Experiment 27 was devoted to two of the basic properties of MOSFET logic inverters - the voltage transfer function and the power dissipation. This experiment deals with the third basic property - the propagation delay, or the time required by the inverter to respond to a change in the input level.

Introduction

The time required for a MOSFET circuit to respond to a changing input signal is determined largely by two factors - the transit time of charge carriers through the devices, and the time required to charge or discharge capacitances in the circuit. The former is usually much shorter than the latter, especially in state-of-the-art IC chips that use submicron-dimension MOSFETs. Therefore the response time is determined mainly by the capacitances, which are of two types - the intrinsic device capacitances (gate oxide and junction capacitances) and parasitic capacitances introduced by the device interconnections.

The <u>propagation delay</u> t_p of a logic gate is defined by

$$t_p = \frac{t_{PHL} + t_{PLH}}{2}, \qquad (28\text{-}1)$$

where t_{PHL} and t_{PLH} are as defined in Fig. 28-1. Assuming a square input pulse, t_{PHL} is the interval between the rise of the input pulse and the fall of the output from V_{OH} to the midpoint between V_{OH} and V_{OL}. Similarly, t_{PLH} is the interval between the fall of the input pulse and the rise of the output from V_{OL} to the midpoint between V_{OL} and V_{OH}.

The propagation delay of the CMOS inverter shown in Fig. 28-2 will be measured in this experiment, and compared with a calculated value. Although the two intervals t_{PHL} and t_{PLH} can be calculated in principle from the equation

$$i_D = C \frac{dv_{OUT}}{dt}, \qquad (28\text{-}2)$$

the current through the inverter is not a simple function of time. A reasonable approximation can be obtained by assuming a constant current that is the average of the initial and final values.

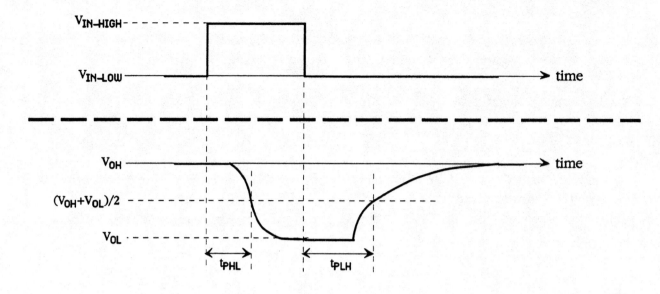

Fig. 28-1

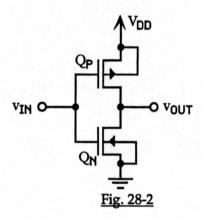

Fig. 28-2

Because $V_{OH} = V_{DD}$ and $V_{OL} = 0$ in the CMOS inverter, and because of the symmetrical voltage-transfer characteristic, (28-2) becomes

$$t_{PHL} = t_{PLH} \approx \frac{C_L \frac{V_{OH} - V_{OL}}{2}}{i_{ave}} = \frac{C_L V_{DD}}{i_{start} + i_{end}}. \qquad (28\text{-}3)$$

The load capacitance C_L represents the capacitance of the following stage in the logic circuit, plus any parasitic interconnection capacitances. The two currents are found in the following way. For the high-to-low transition of the output, immediately after the input pulse turns on, and before the output begins to fall, the NMOS is in the constant-current region because $v_{IN} = v_{GSN} = v_{OUT} = v_{DSN} = v_{DD}$. Also, the PMOS is cut off because $v_{IN} = V_{DD} = v_{GP} = v_{SP}$. Therefore, $v_{GSP} = 0$. The initial current i_{start} can therefore be calculated from the equation for the constant-current region of the NMOS, where $v_{GS} = V_{DD}$.

$$i_{start} = K(V_{DD} - V_{TR})^2 \qquad (28\text{-}4)$$

The load capacitance now begins to discharge through the NMOS at a rate determined by this current. When the output reaches the 50-percent point, the PMOS is still cut off; v_{GSN} is still V_{DD}, but v_{DSN} ($=v_{OUT}$) is now $V_{DD}/2$. The threshold voltage is such that the NMOS is now in the triode region, and the final current i_{end} is calculated from the equation for that region.

$$i_{end} = K\left[2(V_{DD} - V_{TR})\frac{V_{DD}}{2} - \left(\frac{V_{DD}}{2}\right)^2\right]. \tag{28-5}$$

The calculation for the low-to-high transition follows the same route, except that the NMOS is now cut off, and the PMOS supplies the charging current for the load.

Procedure

1) Connect a PMOS and an NMOS on the CD4007 chip (Fig. 28-3) as an inverter, as shown in Fig. 28-2. <u>Recall that the pins of a DIP package are arranged in counterclockwise order when viewed from the top, with pin 1 at the lower left when the indentation is at the left.</u> (See Appendix C.) Be sure that each device has either an internal or an external source-to-substrate connection. Set the power supply to +5 V, and connect a 10-nF capacitor between v_{OUT} and ground. Drive the input with a 0-to-5-V square wave set to a frequency of 10 kHz. Trigger the oscilloscope with the input signal, adjusting the slope control so that triggering occurs on the falling edge of the square wave.

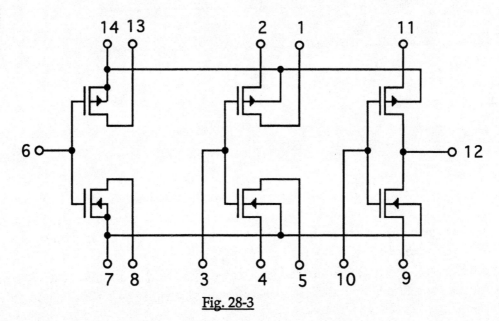

Fig. 28-3

2) Observe the input signal on one channel of the scope, and the output on the other. You may find it helpful to operate the scope in CHOP mode. Determine the 50-percent points of the rising and falling output waveforms, and calculate t_p.

3) If the 10-nF load capacitor is disconnected, the load to the inverter will consist solely of the on-chip capacitances and the parasitic capacitance of the external wiring. The transition times are very small under this condition. Attempt to measure the 50-percent points without the load.

4) An alternative method for measuring propagation delay uses a ring oscillator such as the three-stage version shown in Fig. 28-4. Note that there is no input signal; the output is fed back to the input node.

To analyze the behavior of the ring oscillator, assume the output to be either high or low. By tracing through each of the inverters in turn, you will find that the output is forced to change to the opposite state. The circuit is therefore perpetually unstable, and the output will be a square wave, oscillating between V_{DD} and zero, with the oscillation frequency f depending on the time required for the output signal to propagate through the stages to reach the output node again. The time during which the output remains in one state is one-half the period of oscillation T, and equals the sum of the propagation delays of each stage. Therefore, for the general case of n stages, when n is an odd integer, we have

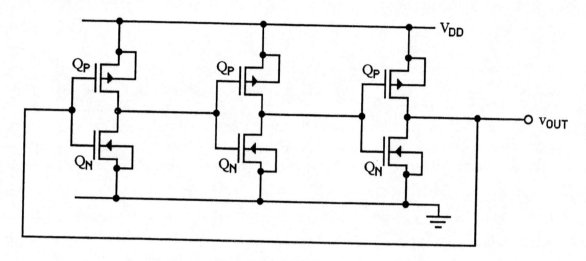

Fig. 28-4

$$t_p = \frac{T}{2n} = \frac{1}{2nf}. \qquad (28\text{-}6)$$

This provides an accurate way for you to measure t_p if n is sufficiently large to allow an accurate measurement of T on the oscilloscope. If you use the CD4007 chip you will be limited to a 3-stage oscillator, unless you can purchase or borrow additional chips to increase the value of n. An alternative is to use one or more of the TC4049BP chips shown in Fig. 28-5, which contains six pairs of CMOS devices, each internally connected as an inverter. Build the oscillator with as many stages as you can, and find t_p by this method. Compare your value with the one you obtained in part 2.

What would happen if you used an even number of stages in your oscillator? Test your conclusion by removing one stage and measuring the output.

5) In a variation on the ring oscillator method, the output of a multistage inverter string (with no feedback loop) is fed to one channel of the oscilloscope, and the input signal is fed to the other, triggering the scope with the input signal. The interval between the rise, or fall, of each

signal, divided by the number of stages, equals t_p. As before, the accuracy of the measurement improves with the number of stages. You might wish to compare the results of the two methods, using either of the CMOS chips.

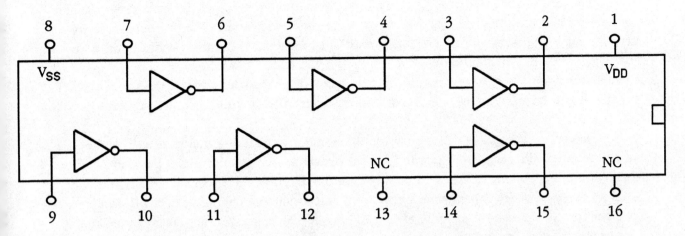

Fig. 28-5

MAJOR DESIGN PROJECTS

The purpose of these projects is to allow you to apply the knowledge you have gained in your circuit theory and electronics courses to design, construct, and evaluate practical electronic circuits. As in all practical situations, the circuit must meet performance specifications, which are given in each case. You should treat the design as if you were a practicing engineer who has been given a work assignment. Your solution should be based on the soundness of the relevant engineering principles, the realism of cost, and the practicality of building and testing the circuit.

Success in these tasks will require planning that may include a literature search, the consideration of alternative designs, and the performance of calculations to determine approximate component values. Although you cannot be sure of a successful design until you build and test it, you must have your preliminary ideas well worked out before you start construction. Otherwise the job will take much longer. The longest of these projects will require four or five weeks to complete.

This is an opportunity to have some fun trying out your ideas. An elegant yet practical design can be a source of great personal satisfaction. But remember that elegance does not necessarily imply complexity. Often the most elegant designs are the simplest ones.

You may use a published circuit if it conforms with the specifications, but be prepared to explain in detail the operation of the circuit and the purpose of every component in it. You may not use special-purpose integrated circuits, such as audio amplifiers, since the goal of the project is for you to design your own circuits.

DESIGN PROJECT 1

FUNCTION GENERATOR, OR MULTIPLE-WAVEFORM GENERATOR

Specifications

You are to design, build, and demonstrate a function generator, or multiple-waveform generator, that meets the following specifications:

1. The generator will provide the following outputs, each at a separate terminal: symmetrical sinusoid, symmetrical square wave, unipolar triangular wave, unipolar pulse, dc, and ramp.

2. A power amplifier will be constructed that, when connected to the appropriate output of the generator, will deliver a signal of the corresponding waveform with minimal distortion, with high efficiency, and with an amplitude controllable between 0 and at least 2 volts peak, to a 10-ohm resistive load. The power amplifier stage can be connected to the generator outputs with a multi-terminal selector switch if you wish, but this is not necessary. Manual connection and disconnection of wires will be acceptable.

3. The frequency of all the periodic waveforms must be adjustable over the range of at least 0.3 kHz to 3 kHz. A single frequency control for all waveforms is acceptable.

4. The amplitude of all outputs must be controllable. A single control for all outputs is acceptable.

5. The ramp output must increase linearly with time over an interval of at least ten seconds. Since it obviously cannot increase without limit, some means for resetting it will be needed. Although simply disconnecting a wire will be sufficient, a more elegant means will merit extra credit.

6. The pulse output may be periodic. There is no requirement for adjustability in pulse width or duty cycle, or for the generation of a single pulse.

7. The generator must be constructed only with components found in your lab kit, with the following exceptions:

   switches and special-purpose passive components, such as multiple potentiometers, may be purchased if desired;

   passive components of any value or any power rating may be used;

8. Special-purpose integrated circuits, other than those in your lab kit, are not permitted.

Guidelines

A suggested block diagram of the generator is shown below. The use of a master oscillator is one approach, but not the only one. Any workable design that you can explain is acceptable. All dc power can be obtained from a lab-bench power supply, but this supply may not be used directly for any of the output waveforms. Note the location of the zero-volt baselines of the output waveforms.

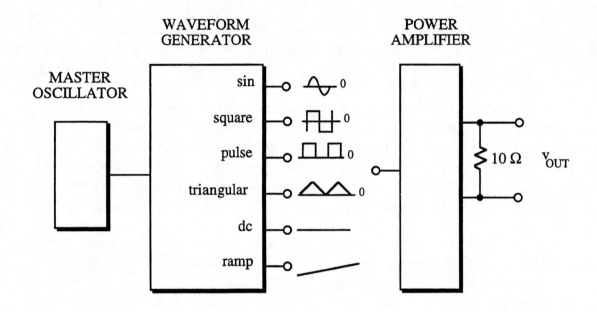

Extra-Credit Work

If you meet all the above requirements, you may earn full credit. However, you may receive additional credit if your generator has any of the following additional features:

1. Provision for controlling the duty cycle of the square wave and/or the width of the pulses in the pulse generator.

2. Control of the symmetry of the triangular wave (unequal rise and fall times).

3. Control of the output amplitude by electronic means, such as an external voltage source, rather than by a potentiometer or variable resistor at the output

4. Any other non-trivial feature.

DESIGN PROJECT 2

AUDIO-FREQUENCY SPECTRUM ANALYZER

Background

A spectrum analyzer measures the amplitude of each of the frequency components contained in an arbitrary waveform. The result is usually displayed on an oscilloscope as a graph of the peak amplitude of each frequency component vs frequency. For a periodic waveform the heights of the peaks are the coefficients of the terms in the Fourier series representation of the waveform. Although a complete analysis of a waveform requires an unlimited frequency range, you will have to restrict yourselves to a small part of the audio-frequency range, partly because of the limitations of operational amplifiers, and partly because you will be designing relatively simple circuits with limited capabilities.

Specifications

You are to design, build, and demonstrate a spectrum analyzer that can display on an oscilloscope a plot of peak amplitude vs frequency over some portion of the audio frequency range. The exact range is not specified; it will be determined by your design. The wider the range, the higher your grade will be. In any case it must be sufficiently wide to demonstrate that your design performs its basic function properly. The analyzer is to be demonstrated with several kinds of input signal, including, as a minimum, sinusoids of different frequency, square waves, and triangular waves.

Ideally, the analyzer should be capable of distinguishing closely spaced frequencies from each other. No specification is placed on this selectivity feature because it is recognized that relatively simple designs cannot have extremely high selectivity. This is another instance in which your best effort, within reason, is encouraged, and will be rewarded.

Guidelines

As is true for any system, there are many valid design approaches. Original designs that work well are always encouraged. Two possible approaches are presented here to stimulate your thinking, but not necessarily to stifle your creativity.

The most straightforward approach, but not necesarily the easiest, is a tunable bandpass filter whose output is directed to the vertical input of the oscilloscope. The horizontal input of the oscilloscope is synchronized with the tuning control of the filter. It must be calibrated in some way so the frequency components can be identified. Calibration of the vertical scale is not necessary, because only the relative amplitudes of the components need be known. A simplified block diagram of this approach is shown below.

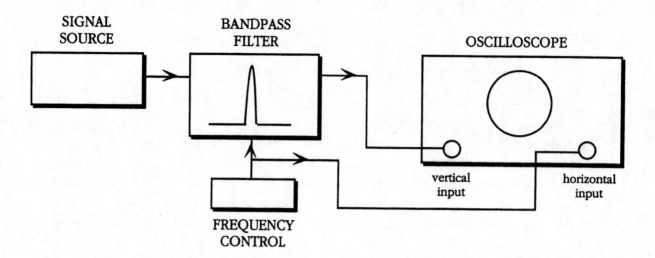

The frequency control may be as simple as a hand-operated potentiometer, or some kind of voltage control of both the filter and the oscilloscope sweep may be devised. Consideration should be given to the question of whether some kind of peak detection of the filter output is required. This method may be difficult for you to realize because tuning a bandpass filter over a reasonably large frequency range is not easy.

The design approach more commonly used in commercial instruments is shown below.

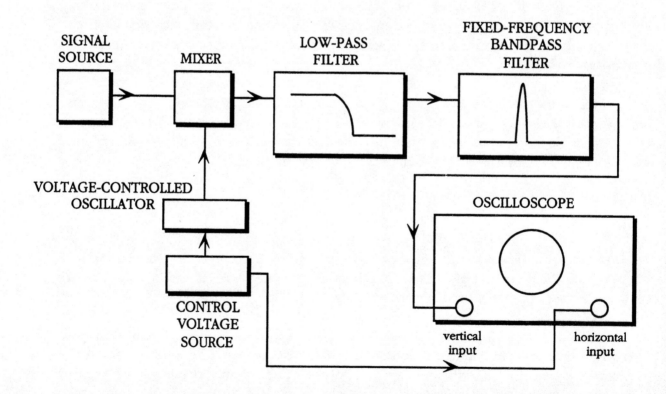

This system is similar to the superheterodyne radio receiver. The signal and the output of a local oscillator are "mixed" to give an output that has frequency components at the sum and the difference of the signal and the oscillator frequency f_0 (as well as other frequencies). The low-pass filter removes the sum frequency, and the bandpass filter transmits the difference frequency f_d. At any time, the output of the bandpass filter represents the component of the input signal whose frequency is $f_0 - f_d$. <u>You may find that a different low-pass filtering scheme is needed.</u>

Because a voltage-controlled oscillator may be difficult to design and build, you may use a commercial function generator that accepts a voltage input to control its output frequency. If you do this, then you will need a second generator to produce your input signals.

Although this approach appears to be more complex than the first one, it may be easier, and more interesting to implement, partly because a bandpass filter is difficult to tune, and partly because signal mixing is a phenomenon worthy of study.

Mixing of signals to produce the sum and difference frequencies is easily accomplished in many ways. Here are several possibilities:

a) multiplication of signals of different frequencies, using the trigonometric relation

$$(\cos\omega_1 t)(\cos\omega_2 t) = 0.5\cos(\omega_1 + \omega_2)t + 0.5\cos(\omega_1 - \omega_2)t$$

b) performing any nonlinear operation on the sum of the two signals - for example, if the two signals are added, then applied to the gate of a MOSFET (a square-law device), the resulting current is proportional to

$$(\cos\omega_1 t + \cos\omega_2 t)^2 = 2 + 0.5\cos 2\omega_1 t + 0.5\cos 2\omega_2 t + \cos(\omega_1 + \omega_2)t + \cos(\omega_1 - \omega_2)t$$

or, if the sum is applied to a diode, there will also be a squared term in the current expression, because the exponential i-v relationship can be expanded in a power series

$$e^x = 1 + x + \frac{x^2}{2!} + \frac{x^3}{3!} + \ldots\ldots$$

c) applying one signal to the input of an amplifier and the other to control the amplifier's bias current. This one may seem strange, but it works! If you use it, you have to explain it.

DESIGN PROJECT 3

AUDIO AMPLIFIER

Specifications

You are to design, build, and characterize an audio amplifier, the magnitude of whose gain is controlled by a dc voltage. The following specifications must be met:

1. The only power source available for your amplifier is a 12-volt, 1 ampere, 60-Hz transformer, with or without center tap, depending upon which is available. The amplifier will require a regulated power supply that you design.

2. The amplifier must operate under two sets of conditions:

   a) An input signal of $0.01\sin\omega t$ volts, to be supplied by a signal generator, operated at or near full-scale amplitude, followed by an appropriate attenuator network;

   b) A voice input, using an inexpensive microphone, or an inexpensive 2-inch speaker operating as a microphone.

3. When responding to the sinusoidal input signal, the amplifier must supply a sinusoidal output signal with minimal distortion, and with a minimum amplitude of 2 volts peak at full gain, to a 10-ohm resistive load with appropriate power rating. The gain should be constant over the frequency range of about 0.1 kHz to 10 kHz.

4. When operating with voice input, the amplifier must provide an easily understandable output from an 8-ohm speaker.

5. The gain of the amplifier must be capable of adjustment by a dc voltage. The gain must change from near zero to the maximum usable value when the control voltage is manually changed from 0 to about +2 volts. Usable gain implies minimal distortion in the output waveform. The gain must be an approximately linear function of the control voltage over a significant portion of the 2-volt range of the control voltage.

A block diagram of the amplifier is shown below:

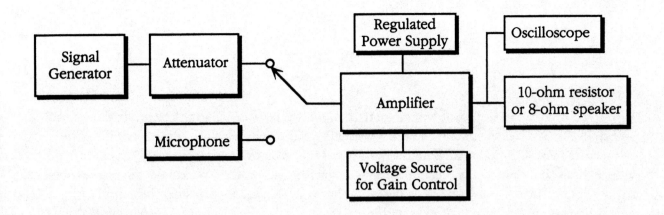

APPENDIX A

USE OF THE BREADBOARD

Your circuits will probably be mounted on a breadboard such as the ceramic "Superstrip" shown in Fig. A-1. Each small rectangle represents a hole into which the component leads and connecting wires fit snugly. Interconnections between the holes are shown by the horizontal and vertical lines that connect the rectangles. The recommended practice is to make the positive power supply connections to one or more of the upper horizontal lines, and the negative supply connections or ground to one or more of the lower horizontal lines. The two rectangular arrays are used for mounting the circuit components.

Your circuit should be arranged in a rectangular pattern that closely resembles the schematic diagram, with interconnections mounted close to the board and no longer than necessary. This practice will simplify the placement of test probes during measurement and debugging.

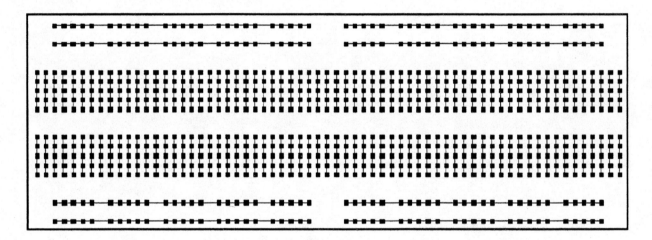

Fig. A-1

APPENDIX B

RESISTOR COLOR CODES AND STANDARD VALUES

Resistors are characterized by their nominal resistance value, tolerance, wattage (heat-dissipation capability), and construction (carbon, metal film, wire-wound). For most carbon and metal-film resistors the nominal resistance value (the intended value) and tolerance (the maximum percentage by which rhe actual value will deviate from the intended value) are indicated by a series of colored bands surrounding the resistor body, as shown in Fig. B-1. The Resistor Color Code (RCC) of the Electronic Industries Association defines a convention for the color coding. Each band color denotes a numerical digit, with the band closest to the end of the resistor body designated as the first band. For a resistor with 10-percent and lower tolerance, the first and second bands indicate the two significant digits used to specify the resistance value. The third band designates the power-of-ten multiplier that follows the significant digits, and the fourth designates the tolerance (Table B-1).

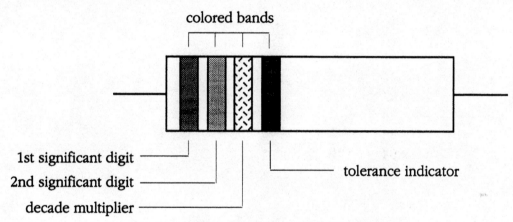

Fig. B-1 Resistor Color-Band Coding System

Table B-1 RESISTOR COLOR CODES

COLOR	AS A DIGIT (1st and 2nd bands)	AS A MULTIPLIER (3rd band)	AS A TOLERANCE (4th band)
Black	0	10^0	1%
Brown	1	10^1	2%
Red	2	10^2	
Orange	3	10^3	
Yellow	4	10^4	
Green	5	10^5	
Blue	6	10^6	
Violet	7	10^7	
Gray	8	10^8	
White	9	10^9	
Gold		10^{-1}	5%
Silver		10^{-2}	10%
empty			20%

Commercial resistors are available in a set of standard nominal values that bridge the entire range of available values. A list of standard significant digits for various tolerance values is shown in Table B-2. Carbon and metal-film resistors are available over the range 10 Ω to 22 MΩ. These standard values and tolerances apply to capacitors as well.

TABLE B-2 STANDARD VALUES AND TOLERANCES FOR RESISTORS AND CAPACITORS

SIGNIFICANT DIGITS	TOLERANCES
10	±5% ±10% ±20%
11	±5%
12	±5% ±10%
13	±5%
15	±5% ±10% ±20%
16	±5%
18	±5% ±10%
20	±5%
22	±5% ±10% ±20%
24	±5%
27	±5% ±10%
30	±5%
33	±5% ±10% ±20%
36	±5%
39	±5% ±10%
43	±5%
47	±5% ±10% ±20%
51	±5%
56	±5% ±10%
62	±5%
68	±5% ±10% ±20%
75	±5%
82	±5% ±10%
91	±5%

The standard values are chosen so that the possible resistance ranges for a given tolerance overlap each other. The actual value of a 5-percent-tolerance 1 kΩ resistor, for example, will lie somewhere between 950 Ω and 1050 Ω. The next lowest 5-percent resistor (910 Ω), will lie between 860 Ω and 955 Ω., which overlaps the low end of the 1-kΩ range.

The novice engineer often makes errors in the specification of resistor values in circuit design. Suppose, for example, that a design calculation calls for a 1.033-kΩ resistor. One might be tempted to connect a 1-kΩ and a 33-Ω resistor in series. Even if their tolerances are only 5 percent, the chance that their sum is exactly 1.033 kΩ is extremely small. If a 1.033-k resistor is truly needed, a more expensive (and much harder to find) resistor with no more than 0.1 percent tolerance would be required.

The uncertainty in component values leads to two important principles that should be followed when designing circuits, and when performing the experiments in this manual.

1) In most cases, circuits should be designed so that they will operate properly when inexpensive 5-percent or 10-percent tolerance resistors are used.

2) When experimental results are being compared with the results of calculations, one must always use the measured values of the components instead of the nominal values in the calculations. Resistors may be measured with an ohmmeter, or calculated from a current vs voltage measurement, using Ohm's Law. If a capacitance meter is not available, capacitances may be calculated from an ac current vs voltage measurement at a known frequency, using the impedance relationship $|Z| = 1/\omega C$.

APPENDIX C

USE OF THE LM741 OPERATIONAL AMPLIFIER AND OTHER INTEGRATED CIRCUITS

Pin diagrams for several versions of the LM741 operational amplifier are shown in Fig. C-1. The most common version is the 8-pin mini DIP (dual inline pin) package. **Pin #1 on DIP packages is identified by the dot on the lid. The pins are ordered countercockwise when viewed from the top, beginning at the lower left when the indentation is at the left side. This convention applies to all integrated circuits, not only to op amps.** If you use an op amp other than the LM741, consult the manufacturer's specification sheet for its connection diagram.

To avoid damage to your op amps, you should observe the following precautions when you assemble and experiment with the circuits that use them.

1. Check the wiring before turning on the dc supply! Make sure the power supply polarity is correct. Incorrect power supply connections are the most common cause of device burnout.

2. Do not apply input voltages that exceed the power supply voltages.

3. Do not apply input signals when the power supply is off.

4. Turn off the power when making changes to the circuit.

5. Do not use unnecessarily long leads.

6. Do not short the op amp output terminal to ground.
(The LM741 can sustain a short circuit at its output terminal indefinitely, but not all op amps can.)

In most experiments, only the following five connections to the op amp are used:

       The positive power supply terminal $+V_{CC}$.
       The negative power supply terminal $-V_{CC}$.
       The -IN inverting input terminal.
       The +IN non-inverting input terminal.
       The output terminal.

The additional terminals, which are used to "fine tune" the op amp performance, are used in the experiments with non-ideal op amps. Most op amps in 8-pin DIP packages use the same pin numbers shown in Fig. C-1 for the five inputs listed above. They differ in the use of pins 1, 5, and 8 because they are not consistent in the functions they perform.

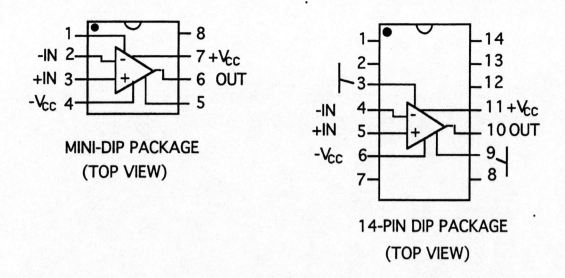

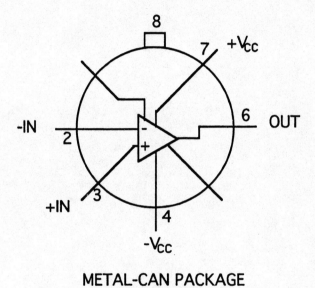

Fig. C-1 Pin Diagrams for the LM 741 Operational Amplifier